Off the Wall Science
A Poster Series Revisited

Author

Harold Silvani

Editors

Betty Cordel
Judith Hillen

Illustrator

Brenda Dahl

AIMS Education Foundation
Fresno, California

This book contains materials developed by the AIMS Education Foundation. **AIMS** (**A**ctivities **I**ntegrating **M**athematics and **S**cience) began in 1981 with a grant from the National Science Foundation. The non-profit AIMS Education Foundation publishes hands-on instructional materials (books and the monthly *AIMS* magazine) that integrate curricular disciplines such as mathematics, science, language arts, and social studies. The Foundation sponsors a national program of professional development through which educators may gain both an understanding of the AIMS philosophy and expertise in teaching by integrated, hands-on methods.

ISBN 1-881431-50-9

Printed in the United States of America

SEAS
AKA82264

Off the Wall Science
A Poster Series Revisited

This book is an accumulation of the activities previously found in the *AIMS Science Posters*. The collection from six sets of posters was compiled into a book in an effort to make them more affordable, thus more accessible to teachers.

The activities were developed to provide students with interesting, inexpensive, easy-to-do, and thought-provoking investigations. In most cases, the materials required to conduct the investigations can be found in the kitchen, garage, or salvaged from the garbage can.

The usage of the activities found in this book is very open-ended. You are invited to adapt them to your needs.

• They can be used at many different levels from awareness to reinforcement of conceptual understanding.

• They can be used as demonstrations, center activities, or with whole-class involvement.

• The activity pages can be copied onto overhead transparencies or enlarged to poster size for whole class viewing, lamenated and posted for use in centers, or used by individual students for designing their own systems of record keeping and accountability.

The activity pages were not designed specifically to require written responses. They pose thought-provoking questions that can be used in discussions or dealt with in journals or science notebooks. Students should always be encouraged to develop their own extensions of the activities and to apply the concepts to their own experiences.

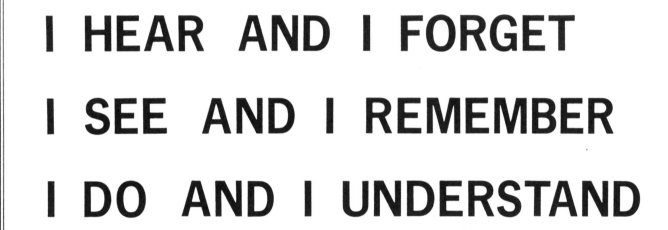

I HEAR AND I FORGET

I SEE AND I REMEMBER

I DO AND I UNDERSTAND

–Chinese Proverb

Table of Contents

The Melting Ice Cube

Topic
Change in the state of matter

Key Question
What causes the ice to change to a liquid?

Focus
Students will observe the change from a solid state of matter to a liquid state.

Guiding Documents
Project 2061 Benchmarks
- *Atoms and molecules are perpetually in motion. Increased temperature means greater average energy of motion, so most substances expand when heated. In solids, the atoms are closely locked in position and can only vibrate. In liquids, the atoms or molecules have higher energy of motion, are more loosely connected, and can slide past one another; some molecules may get enough energy to escape to a gas. In gases, the atoms or molecules have still more energy of motion and are free of one another except during occasional collisions.*
- *Describe and compare things in terms of number, shape, texture, size, weight, color, and motion.*
- *Things change in some ways and stay the same in some ways.*

Science
Physical science
    states of matter

Integrated Processes
Observing
Comparing and contrasting
Collecting and recording data
Interpreting data

Materials
Saucer
Ice cube

Background Information
    Pure water will freeze at 32° F or 0°C. The water crystallizes because the motion of the water molecules becomes so slow that they freeze and become ice. Impurities will lower the temperature at which water will freeze. Because of its salt content, sea water freezes at about 28.5°F. Sugar and alcohol are other substances that lower the freezing point of water in which they are added.
    When water freezes, it increases its volume by 1/11. If 11 cubic inches of water is placed in a freezer, the ice that is formed will take up 12 cubic inches of space. The density of ice is less that the density of the water from which it was formed; this is why ice will float when placed in water. The expansion of ice explains why water pipes break in subfreezing temperatures and why freezing water in an automobile radiator damages a car's engine. Ice starts to melt when the temperature reaches about 32°F.

Properties of Liquids and Solids
Solids:
- will not allow another solid to pass through them easily
- have a definite shape
- have a definite volume
- usually become liquid when heated

Liquids:
- will allow a solid to pass through easily
- take the shape of the container
- have a definite volume
- usually become gas when heated

    More than 21,000,000 tons of commercially manufactured ice are sold in the United States each year. In Canada and in the northern part of the US, some natural ice is cut from rivers and lakes and used in refrigeration processes. The first artificial ice plant was set up in New Orleans in 1868, about three years after the end of the Civil War. The same year, the first refrigerated railroad car was built which enabled many businesses to expand their operations. For example, ranchers in Texas could slaughter their cattle and ship the meat in refrigerated cars to all parts of the United States. Fruit and vegetable farmers in the southern states could ship their products to other markets around the US.

Management
1. When preparing the ice cube, use an ice cube tray, a small plastic container, or an empty butter or margarine tub. Vary the size of the container to accommodate the time period that you will have the students observing the activity.

2. Appoint an ice cube monitor to periodically check the cube. The monitor should be instructed to keep the teacher informed of the progress. During the last few minutes, the class should be involved in making close-up observations.

3. The actual melting time will depend on several things including the size of the piece of ice, placement in the room, and the room temperature.

Procedure

1. Have students estimate the number of minutes that it will take for the ice cube to melt when placed in the saucer.

2. Record the starting time and place the saucer and cube in a place where it can be observed.

3. Have the students calculate the time it will be if their estimated time of melting turns out to be accurate.

4. Direct the students to graph the estimated times. (They will need to determine the range of estimates and set up time intervals on the graph.)

5. Have them determine the average time estimate for the class.

Discussion

1. How long did it take for the ice to melt?

2. Where in the room could the ice be placed so it would melt faster? Why? [It could be placed near a window, near a heater, or on a high shelf because warm air is pushed up by the cooler air and the room temperature would be warmer in a higher position.]

3. What are some of the things that could be done to make the cube melt faster? [Place it in the sunshine, rub it, blow on it, break it into smaller pieces, hold it under hot tap water, etc.]

4. What would the results have been if a larger piece of ice had been used?...a smaller piece? [If other variables were held constant, a large piece of ice would require more melting time and a smaller piece less time.]

5. Compare and contrast the properties of the ice (a solid) with that of the liquid water.

Extensions

1. Have students suggest ways that ice is used in our everyday life. Some suggestions might include the following:
   a. It is used to protect foods from spoiling.
   b. Some sports depend on ice. These include skiing, ice skating, and ice sailing.
   c. It is used to chill beverages and make ice cream.
   d. Ice is used to treat some types of injuries.

2. Prepare several chunks of ice that are different shapes and sizes. Float them in water to observe what amount is below the surface.

3. Partially fill a clear plastic container with water, mark the water level and then place it in a freezer. After the water freezes, remove it from the freezer and note the increased volume taken up by the ice. Is it approximately 1/11 more than the volume taken up by the water? Let the ice melt to find out if the water takes up less volume.

The Melting Ice Cube

Materials:
Saucer
Ice cube

Procedure:
1. Estimate the number of minutes it will take for the ice cube to melt completely.

2. If your estimate is correct, what time will it be when all of the ice has melted?

Questions:
1. How long did it take for the ice to melt?

2. Where in the room could it be placed so it would melt faster? Why?

3. What are some of the things that could be done to make the cube melt faster?

4. What would the results have been if a larger piece of ice had been used?...a smaller piece?

5. What does this activity demonstrate?

Ice Water in a Tin Can

Topic
Condensation

Key Question
What happens to the outside of a tin can when ice water is added?

Focus
Students will observe how cooling affects condensation.

Guiding Document
Project 2061 Benchmarks
- *Raise questions about the world around them and be willing to seek answers to some of them by making careful observations and trying things out.*
- *When liquid water disappears, it turns into a gas (vapor) in the air and can reappear as a liquid when cooled, or as a solid if cooled below the freezing point of water. Clouds and fog are made up of tiny droplets of water.*
- *The cycling of water in and out of the atmosphere plays an important role in determining climatic patterns. Water evaporates from the surface of the earth, rises and cools, condenses into rain or snow, and falls again to the surface.*

Science
Physical science
  water cycle

Integrated Processes
Observing
Comparing and contrasting
Generalizing
Applying

Materials
Water
Ice cubes
2 empty tin cans

Background Information
  Water (H_2O) is the most common substance in the world. It is the only substance on earth that is naturally present in all three forms–solid, liquid and gas. Water covers about 70% of the earth's surface. It is in the ground and in the air we breathe. Our bodies are about two-thirds water. Many scientists believe that life itself began in the water–in the salty water of the sea. Their theory is partially based on the salty taste of a human's blood, sweat, and tears.

  Only about 3% of the water on earth is fresh. About three-fourths of the fresh water is in glaciers and ice caps. On the average, each person in the United States uses about 70 gallons of water each day in the home. Every glass of water that we drink contains molecules of water that have been used countless times before. There is as much water today as there ever was or ever will be. Almost every drop of water we use finds its way back to the ocean where it is evaporated by the sun and then condenses and falls back to the earth as rain, snow, etc.

Management
  Prior to the activity, clean the cans and remove any labels.

Procedure
1. Dry cans thoroughly before adding water.
2. Fill both cans 1/2 to 3/4 with water. Observe the outside surface of cans.
3. Add several ice cubes so that water level in each can is about an inch from the top.
4. Observe outside surface of can for several minutes.

Discussion
1. What happened to the cans after adding water? [Generally, after adding water to the cans, no changes can be observed; however, on a very hot day, a film of water might be seen on the outside surface.]
2. What happened after the ice cubes were added? [As ice cubes are added, the water level inside the can rises.]
3. What do you think would happen if the can was left to stand for an hour?...a day?...a week? [After the ice cubes have been added, beads of water will be seen forming on the exterior of the can. This happens because the ice cools the water, which cools the tin can, which cools the air surrounding it. The water vapor in the air condenses and collects on the surface of the can. After a period of time, the water would warm to room temperature and no condesation would be observed.]

4

4. What does this activity demonstrate? [Condensation is part of the water cycle.]
5. What happens to the water in rain puddles?
6. Why does water have to be added to swimming pools and aquarium periodically?
7. Where does the sweat from our bodies go?

Extensions
1. If a balance is available, find the mass of a pint of water. Compute to determine the mass of a quart...a gallon....a liter. Let the open containers of water sit for a week. Find the mass of the water in each container at the end of the week.
2. Fill a can with water and check it daily to find out how long it takes to completely evaporate. Discuss how the dew forms on things during the night.
3. Suggest to students that they demonstrate this activity at home and discuss the results with their parents. Discuss condensation and evaporation.

Ice Water in a Tin Can

Materials:
Water
Ice cubes
2 empty tin cans
(for best results,
use a tin can with
no paper label)

Procedure:
1. Dry cans thoroughly before filling 3/4 full with water. Observe outside of cans.
2. Add several ice cubes to one can so that water level is about an inch from the top. Observe outside surfaces of both cans for several minutes.

Questions:
1. What happened to the can after putting in water?

2. What happened after the ice cubes were added?

3. What do you think would happen if the cans were left to stand for an hour?...a day?...a week?

4. What does this activity demonstrate?

Water in 5 Containers

Topic
Evaporation, surface area

Key Question
Over a period of time, what will happen to the water in the five containers?

Focus
Students will observe that the larger the surface area of water that is exposed to the air, the greater the evaporation.

Guiding Document
Project 2061 Benchmarks
- *Describe and compare things in terms of number, shape, texture, size, weight, color, and motion.*
- *When liquid water disappears, it turns into a gas (vapor) in the air and can reappear as a liquid when cooled, or as a solid if cooled below the freezing point of water. Clouds and fog are made up of tiny droplets of water.*
- *The cycling of water in and out of the atmosphere plays an important role in determining climatic patterns. Water evaporates from the surface of the earth, rises and cools, condenses into rain or snow, and falls again to the surface.*

Science
Earth science
    water cycle

Integrate Processes
Observing
Comparing and contrasting
Generalizing
Applying

Materials
5 containers to hold water (see *Management*).
Water

Background Information
Water is the most common substance on earth. It is everywhere. It covers more than 70% of the earth's surface. It makes up the oceans, rivers, and lakes and it's even in the air we breathe. Every living thing must have water to live.

Only 3% of the earth's water is fresh water. It is estimated that about three-fourths of the fresh water is frozen in glaciers and icecaps. About one-half of one percent of the water is beneath the earth's surface. Water is the only substance that is naturally present in three different forms; it can be a liquid, a solid, and a gas.

Water moves continually from oceans, lakes, and rivers to the land, and back to the oceans, lakes and rivers. This movement of water is called the *water cycle*. The sun's heat evaporates water from the oceans, rivers, and lakes and it rises as invisible vapor and falls back to the earth as rain, snow, or some other form of water. Most precipitation falls back directly into the oceans. The remainder falls on the rest of the earth. Over a period of time, the water that falls on the earth returns to the oceans and the cycle continues.

There is as much water now as there ever was or ever will be. It seems unbelievable, but the water we drink today could be the same water that some prehistoric sea creatures swam in millions of years ago.

Management
1. Use containers of different shapes and sizes.
2. If desired, have students calculate the area of each container's opening.

Procedure
1. Pour the same amount of water into each container.
2. Place on window sill and observe for several days.

Discussion
1. What happened to the water in the containers after the first day? (Results will vary depending on how much water was placed in each container, the size of the openings, the temperature, etc.)
2. Were any of the containers empty after the first day?...second day?...fifth day? [The results depend on several variables, but generally, the flat, large surfaced pie plate will show signs of faster evaporation than the other containers.]
3. Where did the water go? [The water in the containers changes from a liquid and rises into the atmosphere as a vapor.]
4. What happens to the water on the ground after it rains? [After a rain, much of the water is absorbed into the soil or runs off in streams. The rest is evaporated into the atmosphere.]
5. What does this activity demonstrate? [If conducted properly using several containers with different-sized openings, it will demonstrate that the rate of evaporation depends on the amount of exposed surface area of the container. The larger the surface area exposed to the air, the greater the rate of evaporation.]

Extensions

Several variables can be incorporated in this activity. Encourage the students to repeat the activity with the following adaptations:

1. Add food coloring to the water and observe what remains in the containers after all of the water evaporates.

2. Use two quart jars (or any two containers that are the same size). After adding water to each of the containers, cover one with a piece of glass, waxed paper, plastic wrap, etc. What happens to the water level in each container over a period of time? What would happen if one of the containers were covered with a piece of screen and the other container left uncovered?

3. Have advanced students repeat the activity as suggested. Determine the area of the opening of each of the containers. If all are circular, use the formula πr^2; r=radius; π=3.14. Compare the results to determine if there is a ratio between the surface area and the evaporation rate. Use a graph to show the results.

Water in 5 Containers

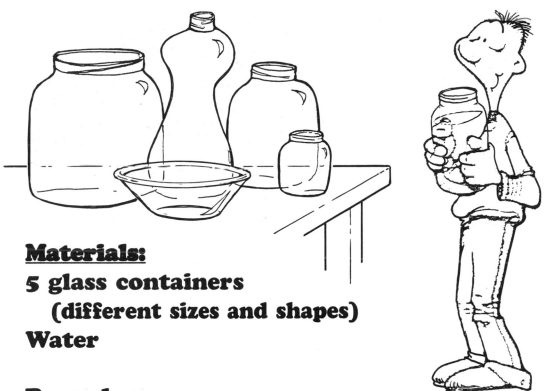

Materials:
5 glass containers
(different sizes and shapes)
Water

Procedure:
1. Pour about a half glass of water into each container.
2. Place on window sill and observe for several days.

Questions:
1. What happened to the water in the containers after the first day?
2. Were any of the containers empty after the first day?...second day?...fifth day?
3. Where did the water go?
4. What happens to the water on the ground after it rains?
5. What does this activity demonstrate?

Salty H₂O

Topic
Evaporation of salt water

Key Question
What happens to salt water when it evaporates?

Focus
Students will observe that salt remains in the pan when salt water is evaporated.

Guiding Document
Project 2061 Benchmarks
- *When liquid water disappears, it turns into a gas (vapor) in the air and can reappear as a liquid when cooled, or as a solid if cooled below the freezing point of water. Clouds and fog are made of tiny droplets of water.*
- *Heating and cooling cause changes in the properties of materials. Many kinds of changes occur faster under hotter conditions.*
- *Seeing how a model works after changes are made to it may suggest how the real thing would work if the same were done to it.*
- *The cycling of water in and out of the atmosphere plays an important role in determining climatic patterns. Water evaporates from the surface of the earth, rises and cools, condenses into rain or snow, and falls again to the surface. The water falling on land collects in rivers and lakes, soil, and porous layers of rock, and much of it flows back into the ocean.*

Science
Life science
   evaporation

Integrated Processes
Observing
Generalizing
Applying

Materials
Salt
Water
Pot
Hot plate or stove
Spoon
Quart jar with lid

Background Information
Salt (NaCl) is a mineral that can be found in either a solid or a liquid form called brine. There is a little more than a quarter of a pound of salt in each gallon of sea water. It is estimated that if all of the oceans dried up, they would leave enough salt to cover all of the United States, except for Alaska and Hawaii, with a layer of salt more than a mile and a half deep. Seas and lakes are salty because streams and rivers flow over the land, dissolving the salt in the soil and rocks, and carrying it to the large bodies of water.

Salt was once so scarce and precious that it was used as money. Julius Caesar's soldiers received part of their pay in common salt. This part of their pay was known as their *salarium*, and it is from this that the word *salary* comes. The modern expression "not worth his salt" originated back in the days when workers were paid all or part of their wages in salt. It literally meant that a man did not earn wages.

Management
Caution: This experiment should be done with adult supervision.

Procedure
1. Mix three tablespoons of salt in one quart of water.
2. Cap the jar and shake vigorously until the salt cannot be seen.
3. Pour the water into the pot and place it on the hot plate. Turn the hot plate on high.
4. Let the water boil until all of it evaporates from the pot.

Discussion
1. Could any of the salt be seen in the water after the jar was shaken? [After shaking the glass, the water will be cloudy because of the salt suspended in it.]
2. What happened to the water in the pot when the hot plate was turned on? [When the hot plate is turned on, the water will begin to heat.]
3. How hot did the water have to get before it started to boil? [Students will probably indicate that the water temperature had to reach 212° F or 100° C before it started to boil. This is true of pure water at sea level, but salt water must get over 212° F before it will boil. This is one reason that salt is added to water when cooking. In addition to flavoring the food, the water retains the heat and the food will cook faster.]
4. Where did the water in the pot go while it was on the

hot plate? [While on the hot plate, the water changes to vapor and evaporates into the air.]
5. Was there anything left in the pot after the water boiled away? [After all of the water evaporates, the only thing that remains is the salt and other impurities that were in the water.]

Extensions
1. During the time that the water is boiling, turn a bowl upside down over the pot so some evaporated water can be trapped. Ask a student to taste it to find out if it is salty.
2. Collect all of the salt from the pot and see if it will fill a tablespoon three times to demonstrate that none of the salt evaporated.
3. Discuss salt content of ocean water. Have students research to find out which is the saltiest body of water in the world. (Dead Sea – 2 1/4 pounds of salt in each gallon of water.)
4. Research and graph the leading salt producing states.
5. Find out the cost of salt per ounce/gram. How much did the salt cost that was used in this activity?

Salty H₂O

Materials:
Salt
Water
Pot
Hot plate or stove
Spoon
Quart jar with lid

Procedure:
1. Mix 3 tablespoons of salt in one quart of water.
2. Cap the jar and shake vigorously until the salt cannot be seen.
3. Pour the water into the pot and place it on the hot plate. Turn on to high.
4. Let the water boil until all of it evaporates from the pot.

Questions:
1. Could any of the salt be seen in the water after the jar was shaken?
2. What happened to the water in the pot when the hot plate was turned on?
3. How hot did the water have to get before it started to boil?
4. Where did the water in the pot go while it was on the hot plate?
5. Was there anything left in the pot after the water boiled away?

Caution: This activity should be done with adult supervison.

Water Drops On a Coin

Topic
Adhesion and cohesion

Key Question
How many drops of water can you fit on the head side of a penny?

Focus
Students will observe the properties of adhesion and cohesion.

Guiding Document
Project 2061 Benchmarks

- *People can often learn about things around them by just observing those things carefully, but sometimes they can learn more by doing something to the things and noting what happens.*
- *Numbers can be used to count things, place them in order, or name them.*
- *Sometimes changing one thing causes changes in something else. In some situations, changing the same thing in the same way usually has the same result.*
- *Describing things as accurately as possible is important in science because it enables people to compare their observations with those of others.*

Science
Physical science
    cohesion
    adhesion

Integrated Processes
Observing
Comparing and contrasting
Applying

Materials
Penny
Water in a cup
Eyedropper
Paper clip
Straw

Background Information
Two important properties of water that are easily observable in classroom science are adhesion and cohesion. The attraction of unlike molecules is called *adhesion* and can be observed by partially filling the glass with water. Close observation will show that the water climbs slightly higher on the walls of the inside of the glass than it does in the middle. This happens because water molecules adhere better to glass than to other water molecules. This can be demonstrated further by turning a glass filled with water upside down. After the water flows from the glass, some drops will remain inside the glass clinging to the sides.

The attraction of one water molecule to another is called *cohesion*. The result of this attraction is called *surface tension* which can be described as an elastic surface or skin that allows some objects to float on it.

The results of the properties of *adhesion* and *cohesion* are more pronounced with cool water. As water molecules cool, the molecular movement of the atoms slows. This results in a tighter, or stronger, attraction of the water molecules and forms a greater surface tension than is possible using warmer water. Some students may have experienced a difference in warm and cold water surface tension when diving from a diving board into a swimming pool. If the water is very cool, the body will experience a stinging sensation as it breaks the surface tension of the water. The stinging sensation is less noticeable in warm water because the attraction of water molecules to each other (cohesion) is less, resulting in less surface tension.

Procedure
1. Place penny on a flat surface so the heads side is facing up.
2. Predict how many drops of water can be put on the penny before it flows off.
3. Put the eyedropper into the cup and draw water up into the stem.
4. Hold the end of the stem about 2 cm above the penny and gently squeeze the rubber bulb so a drop of water is released (one at a time) to fall onto the penny.
5. Continue releasing drops of water until it overflows and runs off the penny.
6. Now flatten one end of a straw by creasing it and folding it over two or three times. Secure this end with a paper clip. Repeat *Procedures 1-5* using the straw as your dropper.

Discussion

1. What happened to the water on the penny when the last drop caused it to overflow? Did only one drop run off the penny? [As the water drops landed on the penny, they began to form a "bubble" of water above the surface of the coin. The property of cohesion holds the water molecules together and the property of adhesion holds the water to the surface of the coin. As the first drop of water flowed off the coin, it pulled more water with it (cohesion).]

2. Do water drops from the straw contain more or less water than the eyedropper? How can you tell? [Experimenting will show that the water drops from a regular drinking straw are larger than those from an eyedropper. (The size of the opening of straws will vary. Select a very small straw and compare the results.)]

Extensions

This activity can lead to several scientific investigations by students in the classroom or at home. Stress the importance of controlling the variables as much as possible. In these investigations some of the variables would include the distance between the eyedropper and the coin, the angle at which the eyedropper is held, the condition of the coin, the temperature of the water, etc. Let students suggest other investigations to add to the ones listed below.

1. Compare the amount of water dropped onto the heads side of the penny with the amount dropped onto the tails side. Repeat this activity three times for each side and then determine total and average for the heads side and the tails side.

2. Use warm tap water (45°- 50° C or 100°-120° F) and compare the number of drops with the amount when cool water (25°-30° C or 70°-80° F) was used.

3. Repeat *Extensions 2* after mixing one tablespoon of salt with the water use same amount of warm and cool water.

4. Using a graduated cylinder, compare the amount of water in a drop from the eyedropper with the amount dropped from the straw. It takes about 100 drops from an average-sized eyedropper or about 60 drops from an average-sized drinking straw to equal 5 ml. At this rate, how many drops from each would it take to fill a 1-liter container?...a half-liter container?

Water Drops On a Coin

Materials:
Penny
Water in a cup
Eyedropper
Paper clip
Drinking straw

Procedure:

1. Place penny on a flat surface so "heads" is facing up.
2. Predict how many drops of water can be put on the penny before the water flows off.
3. Put the eyedropper into the cup and draw water up into the stem.
4. Hold the end of the stem about 2 cm above the penny and gently squeeze the rubber bulb so water is released (one drop at a time) to fall onto the penny.
5. Continue releasing drops of water until it the water runs off the penny.
6. Now flatten one end of the straw by creasing it and folding it over 2 or 3 times. Secure the end with a paper clip.
7. Repeat Procedures 1-5 using the straw as your dropper.

Questions:

1. What happened to the water on the penny when the last drop caused it to overflow? Did only one drop run off the penny?
2. Do water drops from the straw contain more or less water than the eyedropper? How can you tell?

Pepper in the Pie Plate

Topic
Surface tension of water

Key Questions
1. What happens to the pepper when we add soap to the water?
2. How will the results compare for cold water and hot water?

Focus
Students will observe the breaking of the cohesive force of water by soap which has been added to both hot water and cold water.

Guiding Document
Project 2061 Benchmarks
- *Raise questions about the world around them and be willing to seek answers to some of them by making careful observations and trying things out.*
- *Describe and compare things in terms of number, shape, texture, size, weight, color, and motion.*
- *People can often learn about things around them by just observing those things carefully, but sometimes they can learn more by doing something to the things and noting what happens.*
- *Learning means using what one already knows to make sense out of new experiences or information, not just storing the new information in one's head.*

Science
Physical science
    cohesion

Integrated Processes
Observing
Comparing and contrasting
Generalizing
Applying

Materials
2 pie plates
Warm water
Cold water
Pepper
Liquid detergent in a small jar
Toothpick

Background Information
Water is one of the few substances which occur commonly in three different forms on earth–solid, liquid, and gas.

Water molecules are attracted to each other. The attraction of one water molecule to another is called *cohesion*. The result of the attraction, or cohesion, is called *surface tension*. Surface tension causes water to have a "skin" that will support weight. If you carefully lower a needle onto the surface of the water, it will float–the weight is being supported by the "skin," or cohesiveness, of the water molecules. If the needle is dropped onto the surface, it will break the surface tension and sink to the bottom.

Liquid detergent can be used to break the surface tension of water. Soap is used in the cleaning process because it surrounds dirt particles and speeds up the wetting process, thereby making the washing process easier than without soap.

Management
1. If this activity is to be repeated, it will be necessary to thoroughly rinse the pie plates so that no soap residue remains.
2. Make certain that there is a distinct temperature difference in the two water supplies.

Procedure
1. Fill one of the pie plates with warm water, the other with cold water.
2. Sprinkle some pepper onto the water in each of the pie plates and observe the results.
3. Touch the tip of the toothpick to the liquid soap in the small jar.
4. At the edge of the pie plate, slowly immerse the soapy toothpick into pie plate filled with warm water.
5. Repeat *Procedure 3-4* using the pie plate with cold water.

Discussion
1. What happened to the pepper in the warm water when the soapy toothpick was placed near the edge of the pie plate? [When the toothpick with soap on it was placed in the water near the edge of the pie plate, the pepper "scattered" to the other side.]
2. How did the results using cold water compare with the results using the warm water? [The pepper in the cold water did not "scatter" as dramatically as the

pepper in the warm water because the cold water molecules have a stronger bond than the warm water molecules.]

3. How does soap assist in cleaning dishes, clothes, or a person's body? [The purpose of soap in laundering or washing dishes is to emulsify, or to surround, dirt particles with a film. Detergents make washing easier by speeding up the wetting process.]

4. What variables could be added to this activity? [Some of the variables that could be added include the following: (1) compare the results using distilled water with regular tap water, (2) use powdered laundry soap or a bar of soap instead of liquid detergent, (3) sprinkle the pepper on white vinegar or oil and repeat the procedure, and (4) repeat the activity using milk as the liquid.]

5. What is your explanation for what happened? [When soap is added to water, the water's surface tension is broken and the molecules move away from each other.]

Extensions

1. Discuss why soap is added to dishwater or to the water in the washing machine. Have students suggest why dirt is easier to remove when warm or hot water is used with soap. [The soap mixes easier with warm water because the molecules are moving faster than they are in cold water.]

2. Have students bring different brands of soap and test each to find out which one removes dirt the best.

3. To demonstrate how soap speeds up the wetting process, fill two glass tumblers with water. Cut several short pieces of string (2 to 5 cm) and drop an equal number into each tumbler. To one of the tumblers add a teaspoon of liquid detergent and observe. [The string in the tumbler with the soap becomes water-soaked and sinks almost at once. The string in the tumbler without the soap will float on the surface and take a longer time to become water-soaked.]

Pepper in the Pie Plate

Materials:
2 pie plates
Warm water
Cold water
Pepper
Liquid soap in a small jar
Toothpick

Procedure:
1. Fill one of the pie plates with warm water and the other with cold water.
2. Sprinkle some pepper onto the water in each of the pie plates and observe the results.
3. Touch the tip of a toothpick to the liquid soap in the small jar.
4. At the edge of the pie plate, slowly immerse the soapy toothpick into the pie plate filled with warm water.
5. Repeat Procedure 3 and 4 using the pie plate with cold water.

Questions:
1. What happened to the pepper in the warm water when the soapy toothpick was placed near the edge of the pie plate?
2. How did the results using cold water compare with the results using the warm water?
3. How does soap assist in cleaning dishes, clothes, or a person's body?
4. What variables could be added to this activity?
5. What is your explanation for what happened?

THE S.S. PAPER CLIP

Topic
Adhesion and cohesion

Key Questions
1. What happens to the S.S. Paper Clip when we add alcohol to the water?
2. What happens to the S.S. Paper Clip when we add soap to the water?

Focus
Students will observe the results of the reduction of the adhesive attraction when alcohol is added to the water and the reduction of the cohesive attraction of the water when soap is added.

Guiding Documents
Project 2061 Benchmarks
- *People can often learn about things around them by just observing those things carefully, but sometimes they can learn more by doing something to the things and noting what happens.*
- *Describe and compare things in terms of number, shape, texture, size, weight, color, and motion.*
- *Learning means using what one already knows to make sense out of new experiences or information, not just storing the new information in one's head.*

Science
Physical science
    cohesion
    adhesion

Integrated Processes
Observing
Comparing and contrasting
Generalizing
Applying

Materials
Paper clips
2 containers for water
Water
Rubbing alcohol
Liquid detergent
Toothpicks

Background Information
Water molecules are attracted to other water molecules. This attraction is called *cohesion*. *Surface tension* is the result of the attraction between water molecules. Water molecules adhere or cling to other molecules such as glass. This attraction to unlike molecules is called *adhesion*.

Management
1. This activity works best with small paper clips. Try to have different-sized clips to use for comparing results.
2. Use cool tap water.
3. Clear containers such as 9-ounce plastic cups work well. Student groups can use two cups, one for adding the rubbing alcohol and the other for adding the liquid detergent.

Procedure
1. Fill containers (clean and free of any soap) 3/4 full of water.
2. Bend the center part of a paper clip to form a cradle.
3. Use the cradle to lower the other clip very slowly onto the surface of the water. Keep trying until you get the paper clip to float.
4. Observe the floating paper clip from the top and the side.
5. Use the tip of a toothpick to add some rubbing alcohol near the end of the paper clip. Observe what happens to the paper clip.
6. In the second container, float a paper clip and use the tip of another toothpick to add some liquid detergent to the water. Observe what happens to the paper clip.

Discussion
1. Why does the paper clip float? [The paper clip floats because the water molecules cling together and form a "skin" on the surface that supports the weight of the clip.]
2. Describe what the paper clip looks like as it floats. [When the paper clip is floating, the water appears to be climbing onto the metal.]
3. What happens when you add rubbing alcohol to the water's surface? [As alcohol is added to the water, the clip moves away because the adhesion (or attraction of water molecules to the paper clip) is broken.]
4. What happens when you add liquid detergent to the water? [When the liquid detergent is added, the surface tension (cohesion) is reduced and the paper clip sinks to the bottom.]
5. What does this activity prove? [Water can support the weight of certain objects. Things done to water can alter the results.]

Extensions
1. Repeat *Procedures 1* and *2* using hot water. Are the results the same?
2. Add several teaspoons of salt to the water and repeat *Procedures 1* and *2*.
3. Instead of using liquid detergent, use granulated soap...bath soap. Are the results the same?
4. Rub a sewing needle in the palm of your hand. Holding it between your thumb and forefinger, gently lower it to the surface of the water and try to get it to float.

The S.S. Paper Clip

Materials:
Paper clips
Container for water
Water
Alcohol
Liquid detergent
2 toothpicks

Procedure:

1. Use a container that is clean and completely free of soap. Fill 3/4 full of water. Bend the center of a paper clip out to form a cradle.
2. Use the cradle to lower the other clip very slowly onto the surface of the water. Keep trying until you get the paper clip to float.
3. Look carefully at the floating paper clip from the top and the side.
4. Use the end of a toothpick to add some alcohol near the paper clip.
5. Use another toothpick to place one drop of liquid detergent on the water's surface near the paper clip.

Questions:

1. Why does the paper clip float?
2. Describe what the paper clip looks like as it floats.
3. What happens when you add alcohol to the water's surface?
4. What happens when you add liquid detergent to the water?
5. What does this activity prove?

How a Plant Carries Water

Topic
Water transportation in plants

Key Question
How does a plant get water from its roots to its crown?

Focus
Students will observe the xylem vessels in celery.

Guiding Document
Project 2061 Benchmarks
- *Plants and animals both need to take in water, and animals need to take in food. In addition, plants need light.*
- *Animals and plants have a great variety of body parts and internal structures that contribute to their being able to make or find food and reproduce.*

Science
Physical science
Life science

Integrated Processes
Observing
Comparing
Inferring
Applying

Materials
Fresh celery stalks with leaves
Glass of water
Red food coloring
Hand lens, optional

Background Information
Vascular plants have vessels that conduct water and nutrients from the soil through the stems to the leaves. Non-vascular plants, such as mosses, liver worts and horn worts, absorb water directly through their leafy portions.

The roots of most vascular plants are covered by microscopic root hairs which greatly increase the water-absorbing surface of the root. (When transplanting, it is important to leave the roots as undisturbed as possible to avoid tearing off the fragile root hairs. If the root hairs are destroyed in sufficient number, the plant will die because it is unable to absorb the water in the soil.)

From the root hairs, water pases to the *xylem* and flows up the stem to the leaves where much of it is lost (about 90%) by a process known as *transpiration*. Mature corn plants transpire about 30 liters (8 gallons) of water per week. At this rate, an acre of corn may transpire as much as 1,325,000 liters (350,000 gallons) of water in an average growing season which is approximately 90 to 100 days. An average-sized birch tree will transpire from 750 to more than 3,785 liters (200-1000 gallons) of water per day during the growing season. The hotter the day, the more water will be lost. If the human requirement for water were similar to that of plants, each adult would have to drink well over 38 liters (10 gallons) of water per day. Since humans have a closed circulatory (vascular) system, much of the water we consume is recycled.

Plants require a vast amount of water for various reasons. These include: (1) providing a medium for chemical reaction which results in growth, (2) giving structure to non-woody plants, such as the celery plant, and (3) cooling the leaf surfaces during the warm growing season.

The subject of how water moves from the roots to the leaves, against gravity, has been cause for much debate. The most satisfactory explanation of this process, suggested so far, is the cohesion-tension-transpiration theory. The process of water movement through a plant is highly complicated. Simply stated, the transportation of water from the leaf to the external atmosphere causes a force which pulls the water from the xylem to the leaf.

Management
This activity can be conducted by students at home a day or two prior to when celery is going to be included in a family meal. After conducting the investigation, the celery can be used in cooking or eaten raw-the food coloring is safe to eat.

Procedure
1. Use four or five drops of food coloring in a tall glass of water.
2. After cutting off the bottom of the celery stalk, examine the end of the stalk. Use a magnifying glass if available, but even without the aid of a magnifying glass, the xylem can easily be seen between the ridges of the stalk.

3. Place the celery in the glass and mark the water level with a piece of masking tape.
4. The results will begin to show after approximately one hour; however, if the investigaton is started in the morning and resumed a few hours later in the afternoon, students will be able to see more dramatic results.
5. Cut across the celery stalk about 2.5 cm from the bottom. Depending on how long the celery was in the water, you can slice it every 1-2 cm from the bottom and let the students examine the results. Have the students illustrate what they see when examining the celery stalk.
6. Have students bring carnations from home and repeat *Procedures 1-5*.

Discussion

1. When you looked at the bottom of the celery stalk after it had been in the colored water, what did you notice? [Red dots]
2. Explain what you think those red dots are. [Tubes that allow water to go up the stalk of celery.]
3. Did the water make it up to the top of the celery? (Answers will vary upon the time the stalk was in the colored water.) Explain how you know.
4. Break the stalk in half. What are the celery "strings?" [They are the tubes that carry the water up the stalk-the xylem.] How do you know? [They are colored red.]
5. What happened to the water level in the glass during this activity? Why did this happen?

Extensions

1. Have students examine several stalks of celery. Does each stalk have the same number of xylem vessels? Does each stalk have the same number of ridges?
2. Conduct a follow-up investigation by using two glasses of water. Color the water in one glass blue and red in the other glass. Slice the celery stalk up the center about three-fourths of the way to the leaves. Place one-half of the stalk into the glass with the blue water and the other half into the glass with the red water. Leave for several hours and observe. After removing the celery from the glasses, cut across the stalk near the leaves. Do the colors blend or is there evidence of separate red and blue xylem?
3. Select a large celery stalk and find its mass using a balance. Record its mass and then slice it into very thin strips and allow it to dry for several days. Find its mass after the water has evaporated to determine what percent of the celery is water.

How a Plant Carries Water

Materials:
Fresh celery stalks with leaves
Glass of water
Red food coloring
Hand lens, optional

Procedure:
1. Place several drops of red food coloring in a glass of water.
2. Cut off bottom of stalk of celery.
3. Place celery stalk in glass of colored water.
4. After about an hour, remove celery from water.
5. Break celery in middle and observe.
6. Draw a picture of what you see.
7. Try the same activity with white flowers like daisies or carnations. Use different colors of food coloring.

Questions:
1. Compare a fresh stalk of celery to the stalk in the glass of colored water. How are they alike? How are they different?

2. What have you observed about how the stalk of a plant carries water to the leaves?

Weighing objects In Water

Topic
Buoyancy

Key Question
How do the weights of objects compare when weighed in and out of water?

Focus
Students will observe the buoyant force (upward force) that water exerts on objects placed in it.

Guiding Document
Project 2061 Benchmarks
- *Tools such as thermometers, magnifiers, rulers, or balances often give more information about things than can be obtained by just observing things without their help.*
- *Measuring instruments can be used to gather accurate information for making scientific comparisons of objects and events and for designing and constructing things that will work properly.*
- *Tools are used to do things better or more easily and to do some things that could not otherwise be done at all. In technology, tools are used to observe, measure, and make things.*

Science
Physical science
   buoyancy

Integrated Processes
Observing
Comparing and contrasting
Generalizing
Applying

Materials
Spring scale
String
Small jar with lid
Masking tape
Gallon jar or aquarium
Water

Background Information
   Buoyancy is defined as the ability to float or rise in liquid or air. (A buoy is an object floating in water and held in place by an anchor to warn of danger or to mark a channel.) Another definition states that *buoyant force* is the upward force that a liquid or gas exerts on bodies placed in it.

   The Greek philosopher, Archimedes (287-212 B.C.), discovered that the buoyant force which a liquid exerts upon an object in the liquid is equal to the weight of the amount of liquid the object displaces.

   An object under water weighs less than it does out of water because of the water's buoyant force. An object will sink in water if the weight of the water that it displaces is less than the weight of the object. (Example: If a gallon jar is filled to the brim with water and then a 16 pound steel ball is placed into the container, some water would be displaced. This means that some water would overflow and the steel ball would take up the place of the water that spilled out of the container. If all of the overflowed water were collected and weighed, it would weigh less than the steel ball.) On the other hand, if an object is placed in the same container filled to the brim with water and it floats, the weight of the water that overflows is equal to or greater than the weight of the object. (Example: If a piece of wood were placed into the water, some water would spill out of the container. The weight of the spilled water would be equal to or greater than the weight of the wood.)

Procedure
1. Fill the jar or aquarium 3/4 full of water. Fill the small jar 1/2 full of water. Secure the top on the jar. Tie a piece of string around the jar. Tape the string to the glass to keep it from slipping.
2. Using the spring scale, weigh the jar and record the results.
3. Place the jar in the large jar or aquarium while holding it with the spring scale. Record its weight.
4. Remove the lid from the small jar and fill it completely with water. Weigh it out of the water and weigh it in the water.

Discussion
1. Did the jar weigh more in the water or in the air?[The jar when half-filled, and when completely filled with water, weighed more in the air than when in the water. If using a small baby food jar, it will weigh about 200 grams out of the water and just a few grams when placed into the container filled with water. The half-filled baby food jar will float. This means that the water it displaces weighs more than the jar and the water.]

2. **What do you think the** results would be if you used **sand in the small** jar instead of water? (Repeat the procedure using the jar half-filled with sand. Repeat using the jar completely filled.)

3. Does your body weigh more in or out of water? [Some students should be able to relate this principle to the experience of being in the water. Ask if anyone has tried to retrieve an object from the bottom of a swimming pool. A swimmer has to overcome the upward force of the water (buoyancy) in order to get down to the bottom.]

4. What does this activity demonstrate? [This activity demonstrates that objects weigh less in water than they do out of water.]

Extensions

1. Most students have probably experienced the effect of buoyancy on their bodies either in a bathtub or in a swimming pool. You can relate the results of this activity to these experiences. Most people can lie motionless in a swimming pool and the buoyancy of the water will support their weight, causing the body to float on the surface of the water.

2. Add salt to the water and repeat *Procedures 1- 4.* Will the small jar filled with water weigh more or less in the salt water compared to the fresh water?

Weighing objects In Water

Materials:
Spring scale
String
Small jar with lid
Masking tape
Gallon jar or aquarium
Water

Procedure:
1. Fill the jar or aquarium 3/4 full of water. Fill the small jar 1/2 full of water. Secure the top on the small jar. Tie a piece of string around the jar. Tape the string to the glass to keep it from slipping.
2. Using the spring scale, weigh the small jar and record the results.
3. Place the small jar in the large jar or aquarium while holding it with the spring scale. Record its weight.
4. Remove the lid from the small jar and fill it completely with water. Weigh it out of the water and weigh it in the water.

Questions:
1. Did the jar weigh more in the water or in the air?
2. What do you think the results would be if you used sand in the small jar instead of water?
3. Does your body weigh more in or out of water?
4. What does this activity demonstrate?

Aluminum Foil Boats

Topic
Buoyancy, water displacement

Key Question
What design of boat can you build to hold the most pennies?

Focus
Students should observe the objects that are lighter than the water they displace will float.

Guiding Document
Project 2061 Benchmarks

- *There is no perfect design. Designs that are best in one respect (safety or ease of use, for example) may be inferior in other ways (cost or appearance). Usually some features must be sacrificed to get others. How such trade-offs are received depends upon which features are emphasized and which are down-played.*
- *A model of something is different from the real thing but can be used to learn something about the real thing.*
- *Seeing how a model works after changes are made to it may suggest how the real thing would work if the same were done to it.*
- *Use numerical data in describing and comparing objects and events.*
- *Describe and compare things in terms of number, shape, texture, size, weight, color, and motion.*

Science
Physical science
   buoyant force

Integrated Processes
Observing
Comparing and contrasting
Generalizing
Applying

Materials
Piece of aluminum foil (10cm x 10cm)
Aquarium or other container for water
Water
Pennies

Background Information
Most children and probably many adults have wondered what makes a steel ship float. This activity can help to explain about buoyancy and why some things will float and others don't. Sometime during the students' involvement in the activity (it can be either an introductory or closure demonstration), fold the 10cm x 10cm piece of aluminum as many times as possible and press it tightly to remove most of the trapped air. Place it on the surface of the water and observe. It will sink to the bottom. Now remove the foil, unfold it, and smooth it. Gently, place it on the surface of the water and watch it float. If the sides are bent up to form a hull, it will begin to support weight. When a steel ship is built, the hull is designed with "holds" (where cargo is stored) that make air spaces that are needed to make the ship float. The upward force of the water against the bottom of the ship (or the foil boat in this activity) supports the downward pressure of the boat and its cargo.

Anything that floats, including the giant ocean liners, pushes out or displaces as much water as it weighs. As an example, the *Queen Elizabeth* weighs over 83,000 tons so it displaces an amount of water that is equal to 83,000 tons. A classroom demonstration can be used to show this principle. Place a wide-mouth jar in a pie tin. Fill the jar to the top with water. Carefully place an apple into the jar. The water that is displaced will overflow into the pie tin. Pour the water from the pie tin into a pan on one side of a balance. Place the apple in the pan on the opposite side of the balance. The weight of the apple should equal the weight of the water that was displaced.

Students should learn that objects that are lighter than the water they displace will float and objects that are heavier than the water they displace will sink.

Management
1. If conducted as a group activity with several students participating, have them measure the aluminum foil so all have the same size (area) piece.
2. The best procedure is to have pieces cut using a paper cutter and ready for distribution at the beginning of the activity.
3. An aquarium, clear plastic shoe box, a liter box, or a clear glass bowl should be used for the water so students will be able to see the water level change as the pennies are placed on the foil. If other standard weights are available, they can be used.

Procedure

1. Distribute pieces of aluminum foil, pennies, and containers of water.
2. Give students time to design and test various shaped boats.
3. Let students experiment placing the pennies on their boat.
4. Have them use the same shaped boat but encourage them to plan the distribution of the weight of the pennies.

Discussion

1. Did the boat support the same number of pennies the first two times? If not, what might have caused it to support more or less pennies? [Answers will vary. The distribution of the pennies will have an effect on the number that can be supported by the boat.]
2. What happened when the boat was redesigned? Did it support more weight? [Answers will vary. Some will probably discover that the foil can be shaped in a way that is most suitable for supporting a maximum amount of weight.]
3. What happened to the waterline on the outside of the boat as the pennies were added. [As more weight (pennies) is added, the waterline will rise higher on the side of the boat.]
4. If you doubled the size (area) of the foil (from 10cm x 10cm to 10cm x 20cm), do you think that your boat will support twice the number of pennies?

Extensions

1. Discuss the area of the foil used (10cm x 10cm = 100cm^2) and repeat using the same area but different dimensions. Have students try a piece 5 cm x 20cm and/or 4cm x 25 cm (both 100cm^2) and compare results.
2. Double the surface area, such as 10cm x 20cm and repeat *Procedures 1-4*.
3. Pose the question, "What can we do to the water so it will support more weight?" Discussion might lead to the suggestion of adding salt to the water. Select a volunteer to try this at home and report findings to the class.
4. Find the mass of each penny. Determine the greatest number of *grams* that a foil boat supported.
5. Have students design a table to keep record of their data.

Aluminum Foil Boats

Materials:

10cm x 10cm piece of aluminum foil

Aquarium or other
    transparent container

Water

Pennies

Procedure:

1. Cut a 10cm x 10cm piece of aluminum foil and form it like a boat so it will float on the water.
2. Put the boat on the surface of the water and place pennies in it, one at a time, until it sinks.
3. Remove the foil and the pennies from the water. Count the pennies. Repeat Procedure 2.
4. Redesign your boat and repeat Procedure 2 again.

Questions:

1. Did the boat support the same number of pennies the first two times? If not, what might have caused the difference?
2. What happened when the boat was redesigned? Did it support more weight?
3. What happened to the waterline on the outside of the boat as the pennies were added?
4. If you doubled the size (area) of the foil (from 10cm x 10cm to 10cm x 20cm) do you think that your boat will support twice the number of pennies? Explain.

Floating and Sinking

Topic
Density

Key Question
Which objects will float and which will sink?

Focus
Objects with densities less than water's will float while those greater than water's will sink.

Guiding Document
Project 2061 Benchmarks
- *Raise questions about the world around them and be willing to seek answers to some of them by making careful observations and trying things out.*
- *Describe and compare things in terms of number, shape, texture, size, weight, color, and motion.*
- *Offer reasons for their findings and consider reasons suggested by others.*

Science
Physical science
  density
  buoyancy

Integrated Processes
Observing
Comparing and contrasting
Generalizing
Applying

Materials
Large dish or aquarium
Cork
Pencil
Paper clip
Thumbtack
Eraser
Pin
Paper
Cloth
Rubber band
Screw
Clay
Marble
Rubber ball

Background Information
Objects will float or sink depending on the amount of water they displace. The force that seems to hold objects up in water is called *buoyancy*. *Archimedes' principle* says that when a body (object) is placed in fluid (water), the weight it seems to lose is equal to the weight of the fluid that it displaces.

If container A which is filled to the brim with water is placed inside container B which can collect the water that overflows, *Archimedes' principle* can be demonstrated. After submerging an object, such as an apple, (an object that floats) in container A, pour the water that overflowed into container B into a balance pan. Place the apple in the pan on the opposite side of the balance. The pan with the water will be lower because the water has a greater mass than the apple. If the same procedure were followed using an egg (an object that sinks), water that overflowed (displaced) would have less mass than the egg.

In the science laboratory, *specific gravity*, which indicates how the density of one material compares with the density of water, is calculated using distilled water that is 4° Celsius. The above information is generally not necessary for these *Floating and Sinking* activities. However, high achieving students might find it interesting to complete some of the suggestions listed under *Extensions*.

Procedure
1. Have students make predictions before testing objects.
2. Fill the container 3/4 full of water.
3. Try floating each object in water.
4. Make a record of which objects float and which sink.

Discussion
1. Which objects floated?
2. Which objects did not float?
3. What did the floating objects have in common? [Answers will vary. Students will probably indicate that for their size (volume), they were relatively light.]
4. What did the objects that did not float have in common? [Answers will vary. Students will probably indicate that for their size (volume), they were relatively heavy.]
5. What does this activity demonstrate? [This activity demonstrates that light objects usually float and heavy objects usually sink. (This explanation is usually sufficient for primary students; however, some of the following *Extensions* might be applicable.)]

Extensions

1. Discuss and try the following:
   a. What would happen if the clay were flattened and then placed carefully on the surface of the water? [If the ends are bent upward to form a hull, the clay will float.]
   b. Have students bring different kinds of cloth (cotton, silk, wool, etc.) and place them in the water to see what happens.
   c. Try different sizes and kinds of paper. Leave them in the water for several minutes to find out the results.
   d. How could one make the thumbtack float? [Stick it into a cork or eraser.]

2. If a balance is available, place the cork in one pan and then add small paper clips or straight pins to the other pan until both sides are balanced. Place the glass or jar in an aluminum pie tin. Fill the glass with water. Stick a straight pin into the cork. Holding the cork with the pin, immerse it in the water until it is completely submerged. The water flowing from the glass will collect in the pie tin. Pour the water into the pan on the balance opposite the clips or pins. What happens to the balance? [The side with the water is heavier than the clips or pins so it will be lower. If the clips or pins are removed and replaced by the cork, the pans will balance.]

3. Have some of the students research Archimedes and give a report.

Floating and Sinking

Materials:

Large dish or aquarium
Cork
Pencil
Paperclip
Thumbtack   Screw
Eraser   Clay
Pin   Marble
Paper   Rubber ball
Cloth   Rubber band

Procedure:

1. Fill the container 3/4 full of water.
2. Try floating each object in water.
3. Make a record of which objects float and which sink.

Questions:

1. Which objects floated?
2. Which objects did not float?
3. What did the floating objects have in common?
4. What did the objects that did not float have in common?
5. What does this activity demonstrate?

The Orange in Water

Topic
Floating and sinking

Key Question
Does an orange float or sink in water?

Focus
Students will compare the density of an orange to that of water.

Guiding Document
Project 2061 Benchmarks
- *People can often learn about things around them by just observing those things carefully, but sometimes they can learn more by doing something to the things and noting what happens.*
- *Describe and compare things in terms of number, shape, texture, size, weight, color, and motion.*

Science
Physical science
   density

Integrated Processes
Observing
Comparing and contrasting
Generalizing
Applying

Materials
Orange
Deep bowl or small aquarium
Water

Background Information
Some things when placed in water will float while others sink. The upward force of the water against a floating object is called buoyancy. The upward force is always equal to the weight of the object. If a 200 pound log is floating in the water, the upward force against the log is 200 pounds.

All floating objects push out or displace as much water as they weigh. This can be easily demonstrated in the classroom by using a balance, objects that will float, and a small aquarium or a wide-mouth gallon jar. Begin by filling the aquarium or jar to approximately the halfway point. Mark the level of the water with a piece of masking tape. Place an object that will float (small block of wood, half-filled jar of water, etc.) onto the surface of the water. The water will rise above the mark on the masking tape. Scoop the water out of the container and pour it into one of the pans on the balance. Continue removing water until the level is the same as it was before the object was placed in the container. Remove the floating object from the water and place it in the pan on the opposite side of the balance. The object and the water should balance. (There are some variables that make a "perfect balance" almost impossible, but students should develop an understanding of displacement from this demonstration.)

At some point, students should develop the understanding that objects will sink if the water they displace weighs less than the object itself and objects that float displace an amount of water that weighs more than the object. Giant ocean liners displace, or push out, as much water as they weigh.

This activity should be introduced by showing the students several objects and having them predict which ones will sink and which will float. Let them place the objects into the aquarium to find out if their predictions were correct.

Procedure
1. Fill the aquarium about 3/4 full with water. Mark the water level with a piece of masking tape.
2. Before placing the orange in the water, have students predict what will happen.
3. While the orange is in the water, check to see what happened to the water level in the aquarium.
4. Remove the peel from the orange. Be sure to remove as much of the pulp as possible before putting it back into the water.
5. Remove the orange and break it into wedges. Place individual wedge sections back into the water.
6. Hold a large piece of peeling between the thumb and forefinger, put it under water and squeeze. Repeat using another piece of the peeling.

Discussion
1. Did the whole orange float or sink when it was placed in the water? [When the whole orange is placed in the water, it will float.]
2. What part of the whole orange is above the surface of the water? [Approximately 1/5 will be above the surface and 4/5 under the water. This will vary depending upon the thickness of the peel.]
3. Did it float or sink after the peeling was removed?

[If all of the peel and pulp is removed from a ripe orange, the inside of the orange will sink to the bottom. If a half-ripened orange is used, the wedges will have a large amount of air in them and they may float.]

4. Did the small wedges float? (The results will depend upon the use of a ripe or half-ripe orange.)
5. What happened when you squeezed the peeling under water? [When the peeling is squeezed under water, air bubbles will float to the surface.]
6. Using the information observed in *Procedures 1-5*, explain why the orange reacted the way it did. [The orange will float because the peeling contains pockets of air-resembling a life jacket. When the peeling is removed, the inside of a ripe orange will sink because its density is greater than that of water.]

Extensions

1. After marking the water level and placing the orange in the water, measure to find out how many millimeters (convert to centimeters) the water rose in the container. This can be done by using a pen and masking tape to mark the water level before and after the orange was placed in the water. After knowing the number of centimeters that the level of the water increases, we can determine the mass of the orange. If the container of the water is rectangular, calculate: length x width x increase in water level (all in centimeters: 1cm=10mm) to find volume of water displaced. The number of cubic centimeters converts to the same number of grams (1 cubic centimeter = 1 gram). If a cylindrical container, such as a wide-mouth gallon jar is used, calculate: radius^2 x pi(3.14) x increase in water level. Use a balance and gram masses to check your results.
2. Have students estimate the number of wedges in an orange. When oranges are to be served in the cafeteria, bring several into the classroom, peel, and determine number of wedges in each orange. Determine the average number in each. Show results on a graph.

The Orange in Water

Materials:
Orange
Deep bowl or small aquarium

Procedure:
1. Fill the bowl or small aquarium with water.
2. Place the orange in the water to find if it will sink or float.
3. Remove the peel and place the orange back in the water.
4. Break the orange into sections and place each section in the water.
5. Hold a piece of the orange peeling under water and squeeze.

Questions:
1. Did the orange float or sink when it was placed in the water in Procedure 2?
2. Did it float or sink after the peeling was removed?
3. Did the small wedges float?
4. What happened when you squeezed the peeling under water?
5. Using the information observed in Procedures 1-5, explain why the orange reacts the way it did.

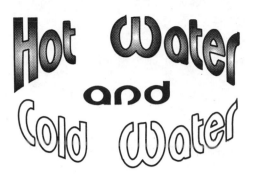

Hot Water and Cold Water

Topic
Densities of hot and cold water

Key Question
What will happen when the hot water meets the cold water?

Focus
Students will observe the mixing of hot water and cold water.

Guiding Document
Project 2061 Benchmarks

- *When warmer things are put with cooler ones, the warm ones lose heat and the cool ones gain it until they are all at the same temperature. A warmer object can warm a cooler one by contact or at a distance.*
- *Seeing how a model works after changes are made to it may suggest how the real thing would work if the same were done to it.*
- *Equal volumes of different substances usually have different weights.*
- *Heat energy carried by ocean currents has a strong influence on climate around the world.*

Science
Physical science
    density

Integrated Processes
Observing
Comparing and contrasting
Applying

Materials
2 identical small glasses or jars
Hot water
Cold water
Food coloring
2 pieces of tagboard

Background Information
A water molecule (H_2O) is made up of two atoms of hydrogen and one atom of oxygen. At standard atmospheric pressure (14.7 lbs. per square inch–PSI) pure water will boil at 100° Celsius (212° Fahrenheit) and will freeze at 0° C (32° F). If the water contains dissolved impurities, the boiling and freezing points will vary slightly.

Water is one of the few substances that occurs commonly in three different forms–solid, liquid, and gas. When it is a solid (ice), it takes up greater amounts of space (volume). When it begins to melt, turning to a liquid (water), it takes up less space; when it changes to a gaseous form, it again takes up more space. As water is heated, the molecules begin to move faster and farther apart. The space between the molecules increases. Equal volumes of cold and hot water do not weigh the same. Because the molecules move faster and are farther apart in hot water, it takes fewer molecules to fill the space in a container; therefore, hot water weighs less than an equal volume of cold water.

The heating and cooling of water has an effect on ocean currents. Water that is warmed near the equator flows away from this area toward the poles as a warm water current. As the warm water flows away from the equator, it begins to cool. As it mixes with cold water far from the equator, it gradually reverses its direction and moves toward the equator as a cold water current.

Management
1. Hot water can be drawn from a tap or heated in a container on a hot plate. It shouldn't be so hot that it will burn the fingers when handled. To make the demonstration more dramatic, use ice cold water in the container that has the cold water.
2. If the glass of cold water is filled so there is a mound of water above the rim of the glass, the card doesn't have to be held when it's turned upside-down. (Secure it to the top of the glass by gently tapping down on the card.)
3. Make sure that the top glass is centered directly over the bottom glass before slowly sliding the card from between the two glasses.

Procedure
1. Fill one glass with hot water and the other with cold water. Put a few drops of food coloring into the cold water.
2. Place a piece of tagboard over the glass of cold water. With one hand, press lightly on the tagboard; with the other hand, pick up the glass and turn it

36

upside down.

3. **Place** the inverted glass of cold water on top of the glass of hot **water**. Be sure that the mouth of the top glass is exactly over the mouth of the bottom glass. Carefully slide the tagboard from between the two glasses.

4. Repeat the above procedures, only this time place the hot water on top and the cold water on bottom.

Discussion

1. What kept the water from falling out of the glass when it was inverted? [The air pressure surrounding the glass and card kept the water from running out. There was more air pressure (weight of the air) pushing up than there was water weight pushing down.]

2. What happened when the tagboard was removed from between the two glasses? [The water in the two containers mixed when the hot water was on the bottom.]

3. Which glass contained hot water and which contained cold water thirty seconds after the completion of *Procedure 3*? [As the hot and cold water mixed after the card was removed, the temperature of the water in both glasses was approximately the same.]

4. What happened when the hot water was on top? [When the hot water was placed on top and the card was removed, the water did not mix. (If observed closely, a slight mixing of the two can be seen where they meet.)]

5. What does this activity show? [This activity demonstrates that hot water is less dense than cold water and that hot water and cold water will mix (form a current) until the temperature equalizes.]

Extensions

1. Place a thermometer in each of the glasses of water. Record the temperature of the cold water and the hot water. What is the difference between the two? How long will it take for the hot water to cool to the point where the temperature in both glasses is the same?

2. Discuss ways that students might observe or experience that cold water is more dense than hot water. Possible responses might include: (1) the water at the bottom of a swimming pool is cooler than the water near the surface; (2) observe what happens when milk or cream is poured into a cup of hot coffee; (3) and watch the reaction when an ice cube is placed into a glass of soda water (if the soda is room temperature).

3. Place a glass of warm, colored water on top of clear cold water, attach a piece of transparent tape around the mouths of the glasses to hold them together. Carefully place them in an out-of-the-way place where they can be observed over a period of time. What happens after a few hours?...a few days? Does the colored water flow down when the temperature equalizes?

4. What would happen if a quart jar of colored hot water were placed over a pint of cold water (both jars need to have the same-sized mouth)?

5. Encourage students to think of other scientific investigations using hot and cold water.

Hot Water and Cold Water

Materials:
2 small jars (same size)
Hot water
Cold water
Food coloring
2 pieces of tagboard

Procedure:
1. Fill one jar with hot water and the other with cold water. Put a few drops of food coloring into the cold water.
2. Place a piece of tagboard over the jar of cold water. With one hand, press lightly on the tagboard; with the other hand, pick up the jar and turn it upside down.
3. Place the inverted jar of cold water on top of the jar of hot water. Be sure that the mouth of the top jar is exactly over the mouth of the bottom jar. Carefully slide the tagboard from between the two jars.
4. Repeat the above procedures, only this time, place the hot water on top and the cold water on the bottom.

Questions:
1. What kept the water from falling out of the jar when it was inverted?
2. What happened when the tagboard was removed from between the two jars?
3. Which jar contained hot water and which contained cold water thirty seconds after the completion of Procedure 3?
4. What happened when the hot water was on top? (Procedure 4)
5. What does this activity show?

Ice Cube in Water

Topic
Relative density of ice compared to liquid water

Key Question
What can you learn from observing ice in water?

Focus
Students will observe that ice will float in water.

Guiding Documents
Project 2061 Benchmarks
- *A model of something is different from the real thing but can be used to learn something about the real thing.*
- *Raise questions about the world around them and be willing to seek answers to some of them by making careful observations and trying things out.*
- *Describe and compare things in terms of number, shape, texture, size, weight, color, and motion.*
- *When warmer things are put with cooler ones, the warm ones lose heat and the cool ones gain it until they are all at the same temperature. A warmer object can warm a cooler one by contact or at a distance.*
- *One way to describe something is to say how it is like something else.*

Science
Physical science
    density

Integrated Processes
Observing
Comparing and contrasting
Generalizing
Applying

Materials
Ice (see *Management*)
Aquarium or other large water container
Water

Background Information
Pure water forms ice at 32°F (0°C). At this temperature, the motion of water molecules becomes so slow that they freeze or crystallize into ice. If the water contains dissolved impurities, the freezing point requires a temperature lower that 32°F or 0°C. For example, because of its salt content, sea water freezes at about 28.5°F.

When water freezes, it increases in volume by 1/11. (This means that when 11 cubic inches of water freezes, 12 cubic inches of ice forms. If allowed to melt, the 12 cubic inches of ice will take up 11 cubic inches when it changes back to liquid water.) This is the reason that ice will float when placed in water; the ice's density (mass to volume ratio) is less than that of the liquid water. (A volume of ice has less mass than the same volume of liquid water.) Glaciers, which have tremendous masses, snowflakes, and particles of frost are all ice; all are less dense than liquid water!

People have used ice for hundreds of years to add to their health and comfort. Before ice was manufactured and sold commercially, natural ice was harvested and stored in icehouses before being shipped to customers. In the United States, natural ice was first transported in 1799, the year George Washington died. At this time, ice blocks were shipped from New York City to Charleston, South Carolina. Speedy clipper ships (fast slender ships with narrow hulls) carried natural ice from New England to many parts of the world. The ice was shipped as far as California during the gold rush and to the East and West Indies, South America, and India.

Ice making machines were being developed around the middle of the nineteenth century. The first artificial ice plant was established in 1868 when Andrew Johnson was the 17th President of the United States. Shortly afterward, refrigerated perishable products were being shipped to distant markets.

Management
1. Prepare different shapes and sizes of ice by using various molds such as an ice cube tray, plastic butter or margarine tubs, plastic bag, small paper cup, etc.
2. A clear glass bowl, plastic shoe or storage box, or a small aquarium can be used for a water container.

Procedure
1. Pour water into a glass jar or aquarium until it is 3/4 full.
2. Place a chunk of ice in the water and observe. (If an ice cube is used, place it so the large end is pointed down.)
3. Repeat with different-sized pieces of ice.

Discussion

1. What happened to the ice when it was placed into water? [The piece of ice floated on the water.]
2. About what fractional part of the ice was above the surface of the water [About 1/11 to 1/8 of the total mass of the ice is above the surface of the water.]
3. Was the result the same when different-sized pieces of ice were used? [When different sizes and shapes of ice are placed in the water, the result is the same. Approximately 1/11 to 1/8 of the total mass is above the surface. (The floating level of the ice will vary slightly depending on the temperature of the water and the amount of impurities in it.)]
4. What would eventually happen if the ice remained in the water for several minutes? [If left in the water for several minutes, the ice will melt completely.]
5. What does this activity demonstrate? [This activity demonstrates that ice will float when placed in water with about 1/11 to 1/8 of the total mass is above the surface.]

Extensions

1. Have students research and give a report on icebergs. Some of the questions that they could be required to answer might include:
   a. How were icebergs formed? [They are pieces of ice that have broken away from a glacier.]
   b. How high do some icebergs tower above the surface of the water? [Some are as much as 400 feet high.]
   c. How many feet below the surface of the water would the bottom of the tallest iceberg be? [About 4,000 feet]
   d. Where are most icebergs located?
   e. What was the Titanic? Give a report on the tragedy.
2. Perform the following scientific investigation to determine if salt water supports weight better than fresh water.
   a. Pour water into two tall glasses so the water level in each is about 8 cm from the top.
   b. Add a couple of tablespoons of salt to one of the glasses of water and stir thoroughly.
   c. Use two ice cubes that are the same size. Place one ice cube in the fresh water and the other in the salt water.
   d. What do you observe? [The ice in the salt water floats slightly higher than the ice in the plain water. The salt adds to the density of the water and supports the weight of the ice.]

Ice Cube in Water

Materials:
Ice
Aquarium or large jar
Water

Procedure:
1. Pour water into a glass jar or aquarium until it is 3/4 full.
2. Place a chunk of ice in the water and observe. If an ice cube is used, place it so the large end is pointing down.
3. Repeat with different-sized pieces of ice.

Questions:
1. What happened to the ice when it was placed into water?

2. About how much (fractional part) of the ice was above the surface of the water?

3. Was the result the same when different-sized pieces of ice were used?

4. What would eventually happen if the ice remained in the water for several minutes?

5. What does this activity demonstrate?

Colored Ice Cubes in Hot Water

Topic
Molecular motion of liquids at different temperatures

Key Question
What observations can be made when a colored ice cube is put in hot water?

Focus
Student will observe that the cold, colored ice water has a greater density than the hot water.

Guiding Documents
Project 2061 Benchmarks
- *A model of something is different from the real thing but can be used to learn something about the real thing.*
- *Raise questions about the world around them and be willing to seek answers to some of them by making careful observations and trying things out.*
- *Describe and compare things in terms of number, shape, texture, size, weight, color, and motion.*
- *When warmer things are put with cooler ones, the warm ones lose heat and the cool ones gain it until they are all at the same temperature. A warmer object can warm a cooler one by contact or at a distance.*
- *One way to describe something is to say how it is like something else.*

Science
Physcial science
  density

Integrated Processes
Observing
Comparing and contrasting
Generalizing
Applying

Materials
Colored ice cube
Hot plate
Pot or tea kettle
Glass container

Background Information
This activity should follow *Ice Cube in Water*. The general information is basically the same; however, it should be remembered that warm and hot water molecules are moving faster and are farther apart than cool and cold water molecules. Because of this, warm and hot water molecules are less dense than cool or cold water molecules. When the colored ice cube is placed in the hot water, it will sink lower into the water than when placed in a container of cool or cold water.

Management
1. A day or two before conducting the activity, add food coloring to some water before pouring it into an ice cube tray.
2. A hot plate is not needed if hot water can be drawn from the tap.
3. An empty quart jar works well for a container for this activity.

Procedure
1. Prior to conducting this activity, have the students discuss what they think the results will be. Ask them to justify their predictions by telling why the results might be as predicted.
2. Heat the water in a pot.
3. After letting the water cool for awhile, pour it into a glass container. (If water is too hot, it might break the glass.)
4. Put a colored ice cube in the container of water.
5. Observe.

Discussion
1. How hot does water have to be before it boils? [At sea level (14.7 pounds per square inch - PSI) distilled (pure) water boils when its temperture reaches 212° F or 100° C. If the water has dissolved impurities in it, it requires a higher temperature before it will boil. At altitudes above sea level, water will boil at temperature less than 212° F (100° C).]
2. How cold must water be before it turns to ice? [Distilled water turns to ice when its temperature reaches 32° F (0° C). Impure water requires a lower temperature before it freezes.]

3. What happens when the ice cube was placed in the water? Why? [When the ice cube was placed in the water, it began to melt. The heat from the water caused the solid (ice) to turn to a liquid (water). The liquid (cold water) was heavier than the warm water around it so it sank to the bottom. The colored ice water could be seen flowing down to the bottom.]
4. What does this activity demonstrate? [This activity demonstrates two things:
   a. Ice is lighter than the same volume of water and it will float when placed in water.
   b When the ice changes to water, the water is cooler (more dense) than the warmer water around it. When warm and cool water mix, the cool water will flow to the bottom.]
5. Apply what you have learned to water in oceans and lakes.

Extensions
1. Ask students to suggest ways that they might observe ice (or cold water) reacting in the way observed in the activity. [Responses might include that ice forms and floats on the surface of frozen lakes or streams. Ice will float when placed in water or beverages. In a swimming pool, the cold water is at the bottom and the warmer water is near the surface. This concept can also be related to warm and cool air. The second floor of a two-story house is warmer than the first floor because warm air rises.]
2. Ask what will happen to the container of water after the ice cube has melted. [The current caused by the mixing of cold and warm water will result in an even distribution after a period of time.]
3. Place a thermometer in the water. How does the temperature of the water compare with the room temperature after one minute?...one hour?...one day?...three days?
4. Have students research to determine at what altitude water will boil when its temperature is 32° F.

The Egg in Water

Topic
Density, salt water and fresh water

Key Question
How can you get an egg to float in water?

Focus
Students will observe that by adding salt to fresh water, the egg will float.

Guiding Documents
Project 2061 Benchmarks
- *Describe and compare things in terms of number, shape, texture, size, weight, color, and motion.*
- *Raise questions about the world around them and be willing to seek answers to some of them by making careful observations and trying things out.*
- *People can often learn about things around them by just observing those things carefully, but sometimes they can learn more by doing something to the things and noting what happens.*
- *Equal volumes of different substances usually have different weights.*

Science
Physical science
    density

Integrated Processes
Observing
Comparing and contrasting
Generalizing
Applying

Materials
Quart jar
Water
Salt
Teaspoon
Egg

Background Information
Salt (NaCl) is a mineral that has more than 14,000 uses. Less than 5 of every 100 pounds produced in the world is used for food seasoning. In cold regions, salt is sometimes used on icy roads. This results in a solution with a lower freezing point than that of water. In some places, salt was once so scarce and precious that it was used as money. Julius Caesar paid his soldiers in common salt. This part of their pay was known as their *salarium*, and it is from this that the word *salary* comes.

The amount of salt in a gallon of sea water ranges from a quarter of a pound in the Atlantic and Pacific Oceans to about two pounds in each gallon of water in the Dead Sea.

This activity demonstrates that when salt is added to water, the solution becomes more dense. When the egg is placed in the fresh water, it will sink; but when it is placed in the salt water solution, its weight is supported by the density of the solution. If just the right amount of salt is added and mixed thoroughly in the water, the egg will float half way between the bottom and the surface of the water in the jar. This must be done carefully by removing the egg each time more salt is added to the water.

Management
1. When measuring the salt, use the flat edge of a knife or ruler to level it in the spoon before adding it to the water.
2. If possible, after removing the egg and adding more salt, cap the jar and shake it to thoroughly mix the solution before replacing the egg. If a lid is not available, stir thoroughly with a spoon.
3. If the egg floats in the non-salted water, you do not have a fresh egg.

Procedure
1. Fill the jar about 5/6 full with fresh, cool water from the tap.
2. Place the egg in water and observe.
3. Remove the egg from the water and put in two level teaspoons of salt. Stir thoroughly with a spoon. Put the egg back in the jar and observe.
4. Repeat *Procedure 3* until the egg floats on the surface of the water.

Discussion
1. What happened when the egg was placed in the fresh water? [When the fresh egg was placed in tap water, it sank.]
2. What was the difference when the egg was placed in the salt water? [The salt water solution will support the weight of the egg better than fresh water.]
3. How many teaspoons of salt did you have to add to make the egg float on the surface?

4. Why do things float better in salt water than in fresh water? [Things float better in salt water than in fresh water because salt water is denser than fresh water.]

5. What do you think might happen to the egg if it were left in the jar for 24 hours?...two days?...a week? [After the egg is floating on the surface of the water, place the experiment on a shelf so the students can observe what happens over a 24 hour, two day, and a week-long.]

Extensions

1. Provide the following information:
   - 28 grams = 1 ounce
   - 1 level teaspoon of salt weighs between five and six grams
   - A large egg will float on the surface of a quart 5/6 filled with water after 9-10 teaspoons of salt are added.

2. Discuss or ask the following quiestions:
   a. About how many level teaspoons of salt would have to be added to a gallon jar 5/6 filled with water in order to float an egg?...a pint jar?... a half gallon jar?
   b. How many grams of salt were added before the egg floated on the surface?
   c. A large egg weighs about 64 grams. How many ounces is this? How many large eggs would it take to weigh a pound?

3. When the egg is floating on the surface, what could be done to make it submerge half way to the bottom of the jar? (Slowly add fresh water to the jar.)

4. Would it be easier to swim in a fresh water lake or ocean water?

5. What might happen to a large frieghter if its cargo is loaded while it's in ocean water and then sails to a fresh water lake to deliver? (The ship's hull will sink lower into the water when it moves from the salt to the same temperature fresh water.)

The Egg in Water

Materials:
Quart jar
Water
Salt
Teaspoon
Egg

Procedure:
1. Fill the jar about 5/6 full with fresh cool water.
2. Place the egg in water and observe.
3. Remove the egg from the water and put in two level teaspoons of salt. Stir thoroughly with a spoon. Put the egg back in the jar and observe.
4. Repeat Procedure 3 until the egg breaks the surface of the water.

Questions:
1. What happened when the egg was placed in the fresh water?
2. What was the difference when the egg was placed in the salt water?
3. How many teaspoons of salt did you have to add to make the egg float to the surface?
4. Why do things float better in salt water than in fresh water?
5. What do you think might happen to the egg if it were left in the jar for 24 hours?...two days?...a week?

Food Coloring in Oil and Water

Topic
Density, surface-tension

Key Question
What happens when we add food coloring to oil and water?

Focus
Students will compare the densities of oil and water and the surface tension of hot and cold water.

Guiding Documents
Project 2061 Benchmarks
- *Describe and compare things in terms of number, shape, texture, size, weight, color, and motion.*
- *Offer reasons for their findings and consider reasons suggested by others.*
- *Equal volumes of different substances usually have different weights.*

Science
Physical science
  density
  cohesion of water

Integrated Processes
Observing
Comparing and contrasting
Generalizing
Appyling

Materials
2 clear soda bottles
Cold water
Hot water
Food coloring
Salad oil

Background Information
Oil is defined as any greasy substance that does not dissolve in water but can be dissolved in ether. The three classes of oil are animal, mineral, or vegetable. They are classified according to their origin. The chief chemical elements of oil are carbon and hydrogen, although some oils contain oxygen. Most oils are less dense than water and are liquid at room temperature.

Oils are classified as "fixed" or volatile according to their behavior when heated. Fixed oils do not easily turn to vapor when heated. They tend to stay in a liquid state.

Fixed oils may be either animal or vegetable. The chief vegetable oils include linseed, cotton seed, soybean, castor, corn and sesame.

Volatile oils turn to gas and evaporate quickly when heated. Volatile oils are used in many ways. They are used in making perfumes, flavoring food and beverages, in making soaps and insect repellents, tobacco, paste, and chewing gum.

Water, the most common substance on earth, is formed by the bonding of two hydrogen atoms and one oxygen atom. A tiny drop of water consists of many millions of molecules, each made up of two hydrogen and one oxygen atom.

Water commonly occurs on earth in three different forms–solid, liquid, and gas. Under standard atmospheric pressure it remains a liquid until its temperature reaches 0˚ C (32˚F). Below this point, it becomes solid-ice . When its temperature reaches 100˚ C (212˚F), it will boil and become a gas-steam. The boiling and freezing points of water will vary slightly depending on the amount of dissolved impurities in it.

The density of water is used as the standard for measuring densities of other substances. As an example, an equal amount of oil is less dense than water. If mixed, the water is more dense, thus heavier for an equal volume, and will sink to the bottom. If water is mixed with an equal volume of molasses, the water is lighter and will rise to the top as the molasses sinks to the bottom.

Two properties of water are adhesion and cohesion. The attraction of unlike molecules is called *adhesion*. If you partially fill a tall, thin glass container with water and observe closely, you will see that the water climbs slightly higher on each side of the container than in the middle. This happens because water molecules adhere better to glass than to other molecules. This is why water drops will remain in a filled glass container when it is turned upside down. Some drops cling to the side of the glass. *Cohesion* is the attraction of one molecule to another of its kind. The result of this attraction is called *surface tension*. Surface tension can be described as an elastic surface, or skin, that allows some objects to float on it.

Management
1. Use two clear bottles with narrow necks (this will allow you to use a small amount of oil) or two jars that are the same size. Two-liter plastic bottles, small olive bottles, test tubes, and graduated cylinders are excellent containers to use.

2. Hot and cold tap water can be used.
3. Regular salad oil (the cheaper the better) and food coloring are necessary.

Procedure
1. Fill one of the bottles with cold water and the other bottle with the same amount of hot water. The water level should be 7 cm below the mouth of the bottle.
2. Pour 2.5 cm salad oil into each bottle. Observe.
3. Add two drops of food coloring to each bottle. Observe.

Discussion
1. What happened when the oil was added to each of the bottles? Why did it happen? [As the oil is added to the water, it will break the surface tension and flow slightly beneath the surface of the water. Air bubbles can be seen clinging to the oil. The oil added to the cold water will rise to the top faster than the oil in the warm water. This happens because the surface tension of the warm water is less than that of the cold water and it flows deeper into the water and disperses more than the oil in the cold water.]
2. What happened in each of the bottles when food coloring was added? How were the results different? [As the colored drops are added to each bottle, they flow slowly down through the oil and float on the surface of the water. Eventually, the colored drop in the warm water breaks the water's surface tension and sinks into the water. The molecular action of the warm water causes the food coloring to disperse. The drop in the cold water will float on the surface for a longer period of time before sinking into the water because the surface tension of cold water is greater than that of warm water. Therefore, it will support an equal amount of weight for a longer period of time.]
3. What do you think the results would be if both bottles were left sitting for an hour?...a day?...a week? (After conducting the activity, let the container set for an hour, a day, a week. Observe the results.)
4. What do you think this activity proves? [This activity shows that oil is less dense than water and will rise to the top when mixed. It also demonstrates that the water's surface tension will support weight.]
5. What other variables could be incorporated in this activity? [Have students discuss some variables that could be incorporated. (Use ice water or very hot water. Use strawberry soda pop instead of the food coloring, etc.)]

Extensions
1. Discuss oil spills in the ocean. Would the oil do more damage if it mixed with the ocean water than it does floating on the surface? (Answers will vary)
2. What can be observed during and after the first rain in the fall? [The oil on the roadways floats to the top of the puddles that are formed. This causes the roads to become slick and dangerous.]
3. What would happen if cold and warm salt water were used?
4. What would happen if the bottles were capped and turned upside down?

49

Food Coloring in Oil and Water

Materials:
2 clear soda bottles
Cold water
Hot water
Food coloring
Salad oil

Procedure:

1. Fill one of the bottles with cold water and the other bottle with the same amount of hot water. The water level should be 7 cm below the mouth of the bottle.
2. Pour 2.5 cm salad oil into each bottle. Observe.
3. Add two drops of food coloring to each bottle. Observe.

Questions:

1. What happened when the oil was added to each of the bottles? Why did it happen?
2. What happened in each of the bottles when food coloring was added? How were the results the same? How were they different?
3. What do you think the results would be if both bottles were left sitting for an hour?...a day?...a week?
4. What do you think this activity proves?
5. What other variables could be incorporated in this activity?

Inverted Tumbler in the Aquarium

Topic
Properties of gases

Key Question
What will happen to the paper towel when we invert the tumbler into the water?

Focus
Air occupies space and can be compressed

Guiding Documents
Project 2061 Benchmarks

- *Air is a substance that surrounds us, takes up space, and whose movement we feel as wind.*
- *Atoms and molecules are perpetually in motion. Increased temperature means greater average energy of motion, so most substances expand when heated. In solids, the atoms are closely locked in position and can only vibrate. In liquids, the atoms or molecules have higher energy of motion, are more loosely connected, and can slide past one another; some molecules may get enough energy to escape to a gas. In gases, the atoms or molecules have still more energy of motion and are free of one another except during occasional collisions.*
- *Raise questions about the world around them and be willing to seek answers to some of them by making careful observations and trying things out.*

Science
Physical science
   matter

Integrated Processes
Observing
Comparing and contrasting
Generalizing
Applying

Materials
Aquarium or wide-mouthed gallon jar
Glass or plastic tumbler
Paper towels
Water

Background Information
Although air has no color, no smell, no taste, and seems like nothing, it is really several gases mixed together. Air cannot usually be seen because it is transparent. Objects can be seen through it because light rays pass straight through air. If there were no air, the world would be silent because most sounds travel through the air.

Air has weight. Although its weight seems relatively little, a paper bag filled with air weighs more than the same bag that is flattened to remove as much air as possible.

The most important gases in air are nitrogen and oxygen. Nitrogen makes up slightly more that 78% of air and oxygen makes up about 21% of the volume of air. The remaining 1% consists almost entirely of the gas called argon. Air contains small amounts of several other gases. They include neon, helium, krypton, xenon, hydrogen, ozone, carbon dioxide, nitrous oxide, and methane.

Air expands when it becomes warm, because of this, warm air can hold more moisture (water vapor) than cool air. Air can be compressed. Machines called compressors can force a large amount of air into small, strong steel cylinders. This compressed air can then be used to do many things, such as inflating tires, providing power to operate pneumatic drills and hammers, and pressurizing the cabins of high-flying airplanes. Compressed air is also used to bring submarines back to the surface.

Management
Students will want to try this themselves, so have several paper towels available to clean up spills and wipe hands.

Procedure
1. Fill the aquarium about 3/4 full of water
2. Before placing the crumpled paper into the glass ask what is in the glass.
3. Crumple a piece of paper towel and push it to the bottom of the glass.
4. When pushing the inverted tumbler to the bottom of the aquarium tell students to observe closely what is happening.
5. Lift the tumbler from the aquarium, turn it upright and remove the crumpled paper towel.

Discussion

1. What was inside the glass tumbler before the paper towel was pushed into it? [Many students might say that the glass was empty; however, it was filled with air.]

2. Did any water enter the glass when it was pushed under water? [Close observation will show that a small amount of water enters the glass as it is pushed to the bottom. This happens because the upward force of the water compresses the trapped air.]

3. Did the paper towel get wet when the glass was pushed under water? [If instructions are followed properly, it will not get wet.]

4. What do you think would happen if you tilted the glass to one side while it is under water? [If the glass is tilted to one side underwater, the air will flow from the glass up to the surface of the water. The air will be replaced with water and the paper towel will get wet.]

5. What does this experiment demonstrate? [This experiment demonstrates that air occupies space and that it can be compressed.]

Extensions

1. Discuss what might happen to the inverted glass if it were pushed down 10 feet into the water. What would happen if it were pushed down into 1,000 feet of water? [The farther down the glass is pushed into the water, the more the air is compressed which results in more water entering the glass. At some point, the glass will probably shatter.]

2. Use a paper cup to demonstrate that air was trapped in the glass. Poke a hole in the bottom of the cup. Invert the cup holding your finger over the hole. Push the cup to the bottom, remove your finger from the hole and observe. [The water will enter the cup forcing the air out the hole in the cup. Bubbles of air will be seen rising to the surface.]

Inverted Tumbler in the Aquarium

Materials:
Aquarium or
  wide-mouthed gallon jar
Glass tumbler
Paper
Water

Procedure:

1. Fill the aquarium about 3/4 full of water.
2. Crumple a piece of paper and push it to the bottom of the tumbler.
3. Hold the tumbler upside-down and slowly push it to the bottom of the aquarium and observe.
4. Lift the tumbler from the aquarium and remove the paper.

Questions:

1. What was inside the glass tumbler before the paper was pushed into it?
2. Did any water enter the glass when it was pushed underwater?
3. Did the paper get wet when the glass was pushed underwater?
4. What do you think will happen if you tilt the glass to one side while it is underwater?
5. What does this activity demonstrate?

The Punctured Plastic Bottle

Topic
Air pressure

Key Question
What will happen when the tape is removed from the hole in the bottle?

Focus
Students will observe the effects of air pressure as it pushes against the water and prevents it from flowing out of the bottle.

Guiding Document
Project 2061 Benchmarks
- *People can often learn about things around them by just observing those things carefully, but sometimes they can learn more by doing something to the things and noting what happens.*
- *Raise questions about the world around them and be willing to seek answers to some of them by making careful observations and trying things out.*
- *Offer reasons for their findings and consider reasons suggested by others.*

Science
Physical science
   properties of air

Integrated Processes
Observing
Comparing and contrasting
Inferring
Applying

Materials
Plastic bottle with screw-type lid
Nail or ice pick
Water
Masking tape

Background Information
This activity demonstrates how air pressure controls the flow of water. When the bottle is filled with water and then capped, the weight of the air around the bottle is exerting a force, or pressure, on it. The force of the air against the hole will "push" against the water and prevent it from flowing out. (As the tape is removed from over the hole in the side of the bottle, a small amount of water might leak from the bottle. This would be caused by some air trapped between the water level inside the bottle and the cap.) When the cap is removed, the air pressure exerts a pressure on the surface of the water equal to the force being exerted against the water in the hole on the side of the bottle. Gravity will then "pull" the water from the bottle.

Air pressure at sea level is approximately 14.7 pounds per square inch (PSI). Air pressure decreases with elevation. At 10,000 feet the air pressure is about 10.2 pounds per square inch. Below sea level, air pressure increases. For example, Death Valley, California is as much as 282 feet below sea level. At that point, the air pressure is greater than 14.7 pounds per square inch.

Management
When presenting this activity to the class, you could poke the hole, place the masking tape and then fill and cap the bottle, before showing it to the students. After removing the tape show the bottle to the class and carefully unscrew the cap while holding it over a waste paper basket. Get the students to think at the analytical/synthesis level by asking them to indicate why the water flows from the hole when the cap is removed and why it stops when the cap is replaced.

Procedure
1. Using a nail or ice pick, make a hole in the side of the bottle about five centimeters from the bottom.
2. Cover the hole with a piece of masking tape and fill the bottle with water.
3. Screw the lid onto the bottle.
4. Remove the masking tape that covers the hole. Observe for several seconds.
5. Unscrew the lid and remove it from the bottle. After five seconds, replace the lid or cover the top of the bottle with your hand. Observe what happens.

Discussion
1. What happened when the masking tape was removed? [When the masking tape was removed none, or very little, of the water leaked from the bottle.]
2. What happened when the lid was removed? What caused this to happen? [When the lid was removed, the water flowed from the hole because of the air pressure pushing on the water through the opening at the top of the bottle and the pull of gravity.]

3. What change took place when the lid was replaced or the opening was covered by your hand? [Replacing the lid, or covering the hole with the hand, cuts off the air pressure on the surface of the water and the flow stops.]
4. What is all around the outside of the bottle? [Air is all around the bottle.]
5. What does this activity prove? [This activity proves that air controls the flow of water.]

Extension

Ask students to suggest ways that air pressure is used in everyday life to control the flow of liquids. Answers will vary, but some responses might include:

a. Use a wedge type can opener to make two holes at the top of a large fruit juice can, tomato sauce can, etc. before pouring. If only one hole is used, the liquid will dribble out.
b. Picnic jugs have an opening at the tip that has to be adjusted before the liquid can be "pushed" out of the spigot.
c. Use a narrow neck vinegar bottle that has the plastic insert in the opening. Fill with either water or vinegar. Hold the bottle upside down and slowly unscrew the cap. The water or vinegar will be held in the bottle by the upward force of the air pressure against the opening.

The Punctured Plastic Bottle

Materials:
Empty plastic bottle with
   screw-type lid
Nail or ice pick
Water
Masking tape

Procedure:
1. Using a nail or ice pick, make a hole in the side of the bottle about five centimeters from the bottom.
2. Cover the hole with a piece of masking tape and fill the bottle with water.
3. Screw the lid onto the bottle.
4. Remove the masking tape that covers the hole. Observe for several seconds.
5. Unscrew the lid and remove it from the bottle. After five seconds, replace the lid or cover the top of the bottle with your hand.

Questions:
1. What happened when the masking tape was removed?
2. What happened when the lid was removed? What caused this to happen?
3. What change took place when the lid was replaced or the opening was covered by your hand?
4. What is all around the outside of the bottle?
5. What does this activity prove?

Card on a Spool

Topic
Air pressure

Key Question
What happens to the card when I blow through the hole in the spool?

Focus
Students will observe that fast moving air results in lower air pressure.

Guiding Document
Project 2061 Benchmarks
- *People can often learn about things around them by just observing those things carefully, but sometimes they can learn more by doing something to the things and noting what happens.*
- *Raise questions about the world around them and be willing to seek answers to some of them by making careful observations and trying things out.*
- *Offer reasons for their findings and consider reasons suggested by others.*

Science
Physical science
    Bernoulli's principle

Integrated Processes
Observing
Generalizing
Applying

Materials
Wooden spool (thread)
5 cm x 5 cm piece of tagboard
Straight pin

Background Information
We do not usually notice the weight of air because it is much lighter than solids or liquids. Air pressure is caused by the weight of air from the top of the atmosphere as it presses down upon the layers of air below. Air pressure at sea level averages about 15 pounds on every square inch. This changes a little from day to day along with the weather. The column of air pressing down on our shoulders weighs about a ton (2,000 pounds). We don't feel this because we are supported by an equal pressure of air from all sides.

Air pressure can be used as a force to lift things. This can be demonstrated by using a straw in a glass of liquid. When the straw is in a glass of milk and the air in the straw is sipped out, the air presses on the surface of the milk in the glass. The milk moves to the path of least resistance which is up into the straw and into the mouth of the person sipping on the straw.

The Swiss Mathematician, Daniel Bernoulli, discovered that the pressure of a gas (air) decreases as the speed of the gas increases. The upper surface of an airplane wing is curved and the bottom surface is almost flat. The curved design causes the air to move faster across the top of the wing than it does beneath the wing. The air pressure is therefore less above the wing than beneath it. The greater pressure from below pushes the plane up giving it lift.

The *Card On A Spool* demonstrates what happens when air pressure is reduced. As the fast moving breath goes through the hole in the spool and out the bottom, it causes the air between the card and spool to move, thereby reducing the air pressure on top of the card. The stable air pressure under the card pushes the card against the spool.

Management
Make sure to remove the paper labels from each end of the spool so that you can blow air through the center hole.

Procedure
1. Stick the pin through the center of the tagboard card.
2. Place the pin and card on top of the spool. The pin should be in the center hole of the spool to keep the card centered.
3. Tilt your head back and hold the spool to your lips. Blow vigorously through the hole for about five seconds.
4. Tilt your head forward, holding the card against the spool. Blow vigorously through the hole for at least five seconds. When the flow of air through the hole has begun, remove your hand from the card.

Discussion
1. Did the flow of air force the card from the spool when your head was tilted back? [If conducted properly, the air passing through the center hole will not force the card off the other end.]

2. Where did the air go that was forced through the hole? [The air passing through the hole moves out over the card.]
3. Describe what happened when the air was forced through the hole while the head was tilted forward. [When blowing through the center hole while the head is tilted forward, the card will cling to the spool until the air flow stops.]
4. Explain the results of this activity. [The air passing through the hole and over the card reduces the air pressure on top of the card. The greater air pressure below the card holds it against the spool.]

Extensions
1. Have students use an encyclopedia to view illustrations showing the effects of air pressure on moving objects (see *Aerodynamics*).
2. A simple experiment that students can do to demonstrate the Bernoulli Principle is to cut a strip of paper about 15 to 20 cm long and about 4 to 5 cm wide. Hold the paper close to the lips so it dangles downward below the chin. Blow vigorously outward above the paper. The fast moving air will reduce the air pressure above it and the stable air pressure below the paper will lift it to a horizontal position.

Materials:
Wooden spool (thread)
5 cm x 5 cm piece of tagboard
Straight pin

Procedure:
1. Make sure to remove the paper labels from each end of the spool so that you can blow air through the center hole.
2. Stick the pin through the center of the tagboard.
3. Place the pin and paper on top of the spool. (The pin should be in the center hole of the spool to keep the card centered.)
4. Tilt your head back and hold the spool to your lips. Blow vigorously up through the hole for about five seconds. Make sure no air escapes around the hole.
5. Tilt your head forward holding the card against the spool. Blow vigorously through the hole for at least five seconds. When the flow of air through the hole has begun, remove your hand from the card.

Questions:
1. Did the flow of air force the card from the spool when your head was tilted back?
2. Where did the air go that was forced through the hole?
3. Describe what happened when the air was forced through the hole while the head was tilted forward.
4. Explain the results of this activity.

Butter and Margarine Candles

Topic
Oil in butter and margarine

Key Question
How will butter and margarine candles vary in their burning times?

Focus
Students will observe that there are oils in butter and margarine.

Guiding Documents
Project 2061 Benchmarks
- *Energy can change from one form to another in living things. Animals get energy from oxidizing their food, releasing some of its energy as heat. Almost all food energy comes originally from sunlight.*
- *Energy can change from one form to another, although in the process some energy is always converted to heat. Some systems transform energy with less loss of heat than others.*
- *Heating and cooling cause changes in the properties of materials. Many kinds of changes occur faster under hotter conditions.*
- *Things that give off light often also give off heat. Heat is produced by mechanical and electrical machines, and any time one thing rubs against something else.*

Science
Physical science
    energy
Health and nutrition
    fats

Integrated Processes
Observing
Comparing and contrasting
Generalizing
Applying

Materials
Butter (1/8 oz.-3 1/2 grams)
Margarine (1/8 oz.-3 1/2 grams)
String
Aluminum foil
Matches
Balance

Background Information
Butter has been used for centuries. The oldest books in the Bible mention the use of butter. The Greeks and Romans used it largely as a medicine. (They preferred olive oil as a cooking fat.)

The United States is the chief butter-producing country in the world. To meet government standards, commercial butter must contain at least 80% fat. It is a valuable energy food with an average of 3410 calories per pound. Butter supplies much of the Vitamin A in the average American diet. If cows are on pasture in the summer, the butter is more yellow and more rich in Vitamin A than in winter.

Butter can be made from sweet or sour cream. Cow's milk is usually used for churning butter, but in some countries it is made from the milk of other animals such as water buffalo, camels, horses, llamas, reindeer, sheep and goats.

A very popular substitute for butter is margarine. It can be produced more cheaply than butter and when Vitamin A is added, it has about the same food value as butter. The fats and oils from such crops as soybeans, cotton seeds, coconuts, peanuts and corn are mixed with water and salt. They are then churned with pasteurized skimmed milk.

There is a long history of laws and taxes assessed to restrict the sale of margarine in packages of one pound or less. Restaurants and boarding houses using colored margarine must clearly identify it. High standards have resulted because of the rivalry between margarine and butter producers.

Procedure
1. Use a balance to determine an equal mass of butter and margarine, each about 3.5 grams.
2. Cut two pieces of string about 5 cm long.
3. Rub the butter and margarine on the pieces of string so each string is saturated with the oils.
4. Mold the butter around one piece of string and the margarine around the other.
5. About 2 cm of the string should be above the mold.
6. Place each candle on a piece of aluminum foil.
7. Use a student assistant to light one of the "candles" while you light the other. Both should begin to burn at the same time.

Discussion

1. How long did each candle burn? Which candle burned for a longer period of time? [Answers will vary. Each candle burns for approximately the same amount of time. Some experiments have shown that each will burn for over 50 minutes.]

2. What is in the butter and margarine that makes them burn like candles? [The oil from the butter and margarine helps to sustain a flame. The heat from the flame melts the solid (butter and margarine) which changes to a liquid and finally to a gas. Through capillary action, the liquid rises in the string. It turns to a gas and burns.]

3. Both butter and margarine are high in calories and supply energy for our bodies. What is the main difference between butter and margarine? (Margarine is produced from vegetable oil and fat. Butter is produced from animal fats.)

Extensions

1. What do you think the results would be if a larger piece of butter and margarine were used?

2. What do you think the results would be if heavier (thicker) string were used?...lighter (thinner) string?

3. Have students complete a research project to learn how the Greeks and Romans used butter. [Both used it as a medicine.]

4. Construct a graph showing leading butter and margarine producing states. Culminate the study of butter by having students make their own butter in the classroom This is done by filling a pint or quart jar half way with cream. Add salt and shake for about 10 minutes. Provide crackers and let the students eat the results.

Butter and Margarine Candles

Materials:
Butter (3.5 grams)
Margarine (3.5 grams)
String
Aluminum foil
Matches

Procedure:
1. Use a balance to determine an equal mass of butter and margarine, each about 3.5 grams.
2. Cut two pieces of string each about 5 cm long.
3. Rub the butter and margarine on the pieces of string so each string is saturated with the oils.
4. Mold the butter around one piece of string and the margarine around the other.
5. About 2 cm of the string should be above the mold.
6. Place each candle on a piece of aluminum foil.
7. Use a student assistant to light both "candles" at the same time.

Questions:
1. How long did each candle burn? Which candle burned for a longer period of time?
2. What is in the butter and margarine that makes them burn like candles?
3. What do you think the results would be if twice as much butter and margarine were used?
4. What do you think the results would be if heavier (thicker) string were used?...lighter (thinner) string?

Caution: This activity should be done with adult supervision.

The Burning Walnut

Topic
Heat energy

Key Question
Which produces more heat energy, a burning walnut shell or a burning walnut meat?

Focus
Walnut oil is fuel for a fire.

Guiding Document
Project 2061 Benchmarks
- *Energy can change from one form to another, although in the process some energy is always converted to heat. Some systems transform energy with less loss of heat than others.*
- *People can often learn about things around them by just observing those things carefully, but sometimes they can learn more by doing something to the things and noting what happens.*
- *Raise questions about the world around them and be willing to seek answers to some of them by making careful observations and trying things out.*

Science
Physical science
    energy

Integrated Processes
Observing
Comparing and contrasting
Generalizing
Applying

Materials
Walnut
2 paper clips
Aluminum foil
Matches
Tape

Background information
There are several types of walnuts. The most common is the English walnut which has a thin shell and fine flavor. There are two kinds of English walnuts, the "Santa Barbara," which grows only along the coastal plains and the valleys of southern California, and the "French," which grows from central California to Oregon. Walnuts, which have much food value, contain fats and proteins. After the nuts are shaken down from the tree, they are hulled and dried. The better grades of nuts are sorted, sized, bleached and sacked for commercial sales. The poorer grades are shelled and used to make walnut oil and shell flour. The United States leads the world in the production of oil, followed by France, Italy, and India. California produces about 80,000 tons per year, followed by Oregon with about 4,000 tons.

Botanically speaking, the walnut is classified as a "drupe." A drupe is a fleshy fruit with a single seed enclosed by a hard, stony shell or pit. The fleshy fruit or nut meat of the walnut contains a large amount of oil. When heated, the walnut oil can sustain a flame much like a candle. When burning, it gives off heat and light energy.

Procedure
1. Remove the shell from a walnut. Try to keep the nut meat in large pieces.
2. Bend the paper clips to construct two stands. Tape the base of each clip to the foil to keep it straight. Attach a large piece of nut meat to one stand and a large piece of shell to the other.
3. Hold a lighted match to each and observe.

Discussion
1. What did you observe about the burning nut meat and shell? [Both the nut meat and shell will burn. Generally, the shell needs more heat to start burning, and it will burn slower than the nut meat.]
2. What kind of energy is produced by the burning material? [The burning material produces heat and light energy.]
3. Where does stored energy in the walnut come from? [The stored energy in the walnut comes from the sun which makes possible the plants' photosynthetic process.]
4. How do we express the amount of energy contained in the food which is taken into the body? [The energy contained in foods is expressed in calories.]
5. What do you think the results would be if the same activity were repeated using a peanut instead of a walnut? (Have students repeat activity using a peanut.)

Extensions
1. Discuss how certain foods are quick sources of energy because of the high caloric content.
2. Conduct an experiment to determine if a walnut or a peanut transfers more caloric energy. This can be done by using four empty tomato sauce cans. Remove both lids from two of the cans. Using a wedge type can opener, make four vent holes on each end of the two cans. These will serve as stands. Remove only one lid from the other two cans. These will be the containers. Attach a piece of walnut meat to a paper clip as shown on activity sheet. Attach an equal mass of peanut meat to a second paper clip. Put about 200 milliliters of water in each of the containers. Put a thermometer in each container and record the temperature. Light the nut meats and place a stand over each. Set the containers with the water and thermometer on each stand. Record temperature one minute after flames go out. Determine the increase in temperature of each container of water.

The Burning Walnut

Materials:
Walnut in the shell
2 Paper clips
Aluminum foil
Matches
Tape

Procedure:

1. Remove the shell from a walnut. Try to keep the nut meat in large pieces.
2. Bend the paper clips to construct two stands. Tape the base of each clip to the foil to keep it straight. Attach a large piece of nut meat to one stand and large piece of shell to the other.
3. Hold a lighted match to each and observe.

Questions:

1. Did either the shell or the walnut meat burn? Which gave off the most heat?
2. What kind of energy is produced by the burning material?
3. Where does stored energy in the walnut come from?
4. How do we express the amount of energy contained in the food which is taken into the body?
5. What do you think the results would be if the same activity were repeated using a peanut instead of a walnut?

Caution: This activity should be done with adult supervision.

The Covered Candle

Topic
Combustion

Key Question
What happens when we place a jar over a burning candle?

Focus
Oxygen supports combustion

Guiding Document
Project 2061 Benchmarks
- *People can often learn about things around them by just observing those things carefully, but sometimes they can learn more by doing something to the things and noting what happens.*
- *Raise questions about the world around them and be willing to seek answers to some of them by making careful observations and trying things out.*

Science
Physical science
   combustion

Integrated Processes
Observing
Generalizing
Applying

Materials
Quart jar
Birthday candle
Matches
Stopwatch or clock with a secondhand

Background Information
Candles were the chief source of light for at least 2,000 years. Over two thousands years ago, crude candles were made of fats wrapped in husks or moss. Most modern candles are made of stearin, which is obtained from tallow, and paraffin, a mineral wax. Wicks are usually made of woven cord.

Candela is a standard for measuring the intensity, or strength of light sources. It was officially adopted in the United States in 1963. A 40-watt bulb gives off 31.36 candelas of light. The light that travels away from a light source is measured in *lumens*. The British standard candle was the first standard of light intensity.

The candle was made from a white crystalline found in sperm whales and was exactly 7/8 of an inch thick. The amount of light given off by the candle was called one candle power. One candela is slightly less than one candle power.

As with anything that burns, candles must have a supply of oxygen in order to sustain a flame. When a burning candle is covered with a quart jar, no more oxygen can get into the jar so the flame will be extinguished when most of the oxygen trapped in the jar is consumed. As the candle burns, carbon dioxide is formed. Because carbon dioxide is heavier than air, it will begin to settle to the bottom of the jar. This will also help to extinguish the flame.

When a candle is burning, the flame is actually burning a gas that is produced by the melting paraffin. The "liquid" wax travels up the wick by capillary action and burns around and above the wick. To demonstrate that the gas, and not the wick, is burning, blow out the candle flame and quickly lower a lighted match to within a quarter of an inch of the wick. The gas will be ignited above the wick.

Management
1. Any kind of candles can be used; however, it is important that the same type (made of the same kind of tallow or paraffin and size of wick) be used when conducting the *Extensions*.
2. When demonstrating this activity to the class, discuss safety procedures.
3. Have students predict what they think will happen. If they indicate that the candle will "go out," have them estimate how long it will burn before going out. Math can be integrated in the lesson by determining the range, median, and mode of the estimates.

Procedure
1. Place the candle on the table and light it. Anchor it in clay or melted wax.
2. Cover the burning candle with the jar or glass.
3. Check the secondhand on the watch for time.
4. Observe.

Discussion
1. What was in the jar when it was placed over the burning candle? [Although many students will say that nothing was in the jar, it was filled with air.]

2. What happened after the jar was placed over the burning candle? Why did it happen? [About 20 seconds after the jar was placed over the candle, the flame was extinguished. This happened because most of the oxygen was consumed in the burning process.]
3. What would the results be if a larger jar had been used to cover the candle? [If a larger jar were used, the candle would burn for a longer period of time.]
4. What do you think the results would have been if a shorter candle had been used?
5. What does this experiment prove?[This experiment demonstrates that something in the air (a gas called oxygen) is needed to sustain a flame.]

Extensions
1. Ask students if they have seen someone blow on a flickering fire to get it to start burning. This is done to move more air toward the flame. It is not the "air" from the person's mouth that feeds more oxygen, because much of the air coming from the mouth is carbon dioxide which will smother the flame. It's the air being moved between the mouth and the flame that helps feed the fire. If possible, show a picture of a bellows. Some students might have seen a movie on TV where a bellows was being used by a blacksmith.
2. Use two candles (same size and length) and a quart and gallon jar. Cover both candles at the same time. How much longer did the candle burn under the gallon jar than the quart jar? (It will not be exactly 4 times as long because of variables such as currents inside the jar, settling of carbon dioxide, etc.)
3. Using the same size candle, cover it with a pint jar. How does the burning time compare with the quart jar?
4. Have students estimate how long a birthday candle will burn (how many minutes) before going out. Record starting time and assign a candle monitor to be in charge of checking the candle periodically. The monitor will determine the time that the flame was extinguished. While the candle is burning, have the students compute and record the time that it will be when the flame goes out. Example: If the candle is started at 9:56 and John predicts 19 minutes, it will be 10:15 when it is extinguished.

The Covered Candle

Materials:
Quart jar
Birthday candle
Matches
Stopwatch or clock with
   a secondhand

Procedure:
1. Place the candle on the table and light it. Anchor it in clay or melted wax.
2. Cover the burning candle with the jar.
3. Check the secondhand on the watch for time.
4. Observe.

Questions:
1. What was in the jar when it was placed over the burning candle?
2. What happened after the jar was placed over the burning candle? Why did it happen?
3. What would the results be if a larger jar had been used to cover the candle?
4. What do you think the results would have been if a shorter candle had been used?
5. What does this activity prove?

**Caution: This activity should be done with
adult supervision.**

Homemade Fire Extinguisher

Topic
Chemical reactions

Key Question
How can we make our own fire extinguisher?

Focus
Students will observe the chemical reaction that occurs when baking soda and viengar are combined. They will observe that the CO_2 gas that is produced does not support combustion.

Guiding Document
Project 2061 Benchmarks
- *People can often learn about things around them by just observing those things carefully, but sometimes they can learn more by doing something to the things and noting what happens.*
- *Raise questions about the world around them and be willing to seek answers to some of them by making careful observations and trying things out.*
- *Offer reasons for their findings and consider reasons suggested by others.*

Science
Physical science
   chemistry

Integrated Processes
Observing
Comparing and contrasting
Generalizing
Applying

Materials
2 empty quart jars
Matches
Small candle
25 cm piece of wire
Baking soda
Spoon
Vinegar

Background Information
   Human beings and animals inhale an air mixture that is about 12% oxygen, 78% nitrogen, and 1% other gases. They exhale carbon dioxide that is produced by *oxidizing* (burning) food in their bodies. Green plants take carbon dioxide from the air and give off oxygen when light shines on them. In the light, plants combine carbon dioxide with water to make food. The air is about 0.035% carbon dioxide.

   Carbon dioxide has everyday uses. It is used to make cakes rise in an oven because baking powder or yeast in the cake batter releases carbon dioxide. It is used to produce the fizz or sparkle in soft drinks, beer, and sparkling wines. It becomes a solid when it is cooled to - 109.3° F (-78.5° C) at atmospheric pressure. The solid is called "dry ice" because it does not melt to form a liquid as water does; instead, it changes directly back into a gas.

   Fire extinguishers commonly used in homes and vehicles spread a dry, snow-like solid that changes to a gas under atmospheric pressure. The gas is carbon dioxide (CO_2) which is odorless, colorless, and tasteless. When it is discharged from the metal container, the carbon dioxide gas settles over a fire and shuts off the supply of oxygen in the air that a fire must have to burn. It does this because it is over one and a half times as dense as air.

   Although most people don't realize it, chemical reactions take place around us constantly. Burning wood in a fireplace, washing clothes in a washing machine, driving a car, breathing, digesting food are daily activities during which chemical reactions occur. The average kitchen has all the materials necessary for many exciting chemistry experiments.

   The gas produced by the *Homemade Fire Extinguisher* is a result of a chemical reaction. When vinegar (acetic acid) is combined with baking soda (sodium bicarbonate), water and carbon dioxide are produced. The gas that is produced in the jar forces the air mixture out. When the burning candle is lowered into the jar, the flame is extinguished because all that remains in the jar is carbon dioxide which doesn't support combustion.

Management
1. When demonstrating this activity, take time to discuss safety precautions, such as lighting the candle, closing the match cover before striking match, etc.
2. After *Procedure 5* extinguish the candle and lower it into the jar two or three times.
3. Wait at least two minutes after mixing the vinegar and baking soda before capping the jar.

Procedure
1. Wrap the wire around the candle
2. Light the candle and slowly lower it into each of the empty jars.

3. Put two spoonfuls of baking soda in one of the jars and 100 ml of vinegar into the other one.
4. Repeat *Procedure 2* by lowering the burning candle into the jar with baking soda and then into the one with vinegar.
5. Pour the vinegar into the jar with the baking soda, wait for about 10 seconds, and then lower the burning candle into the jar.

Discussion

1. What happened when the burning candle was lowered into the jar filled with air?...into the jar with baking soda?...into the jar with vinegar? [The candle continues to burn when lowered into both "empty" jars because there is enough oxygen to sustain the flame. There is oxygen in the air above the vinegar and baking soda.]
2. What happened when the vinegar was mixed with the baking soda? [When the vinegar and baking soda were mixed, carbon dioxide gas was formed. Because carbon dioxide is heavier than air, it remains in the jar and forces out the fresh air. Carbon dioxide does not have the oxygen that is necessary to keep the candle burning.]
3. What happened when the burrning candle was lowered into the mixture? [The CO_2 gas that was produced smothers the flame the same as the gas sprayed from a fire extinguisher.]

Extensions

1. Discuss the use of a portable fire extinguisher and why it is important to have one in the home and in a vehicle.
2. Also discuss what extinguishes a flame from a match when a person blows it out. Is it the force of the wind or the carbon dioxide exhaled by the person?
3. Repeat *Procedure 5*. See how many times a candle can be lowered into the jar and be extinguished before the CO_2 has dissipated.
4. Repeat *Procedure 5* and cover the mouth of the jar with a book. Leave overnight and lower a candle into the jar 24 hours later. Did the carbon dioxide gas extinguish the candle flame?
5. Have one of the students do the activities at home and cap the jar for one week. Remove the cap from the jar and lower a burning candle into it. Was there enough carbon dioxide to extinguish the flame? (Caution students to wait for at least two minutes before capping the jar after mixing the vinegar and baking soda.)

Homemade Fire Extinguisher

Materials:
2 empty quart jars
Matches
Small candle
25 cm piece of wire
Baking soda
Spoon
Vinegar

Procedure:
1. Wrap the wire around the candle.
2. Light the candle and slowly lower it into each of the empty jars.
3. Put 2 spoonfuls of baking soda in one of the jars and 100 ml of vinegar into the other one.
4. Repeat Procedures 1-2 by lowering the burning candle into the jar with baking soda and then into the one with vinegar.
5. Pour the vinegar into the jar with the baking soda, wait for about 10 seconds, and then lower the burning candle into the jar.

Questions:
1. What happened when the burning candle was lowered into the jar filled with air?...the jar with baking soda?...the jar with vinegar?
2. What happened when the vinegar was mixed with the baking soda?
3. What happened when the burning candle was lowered into the mixture?

Pouring Carbon Dioxide Gas

Topic
Chemical reactions

Key Questions
1. What happens when vinegar and baking soda are mixed?
2. What effect does this mixture have on a flame?

Focus
Baking soda mixed with vinegar produces carbon dioxide gas which will extinguish a flame.

Guiding Document
Project 2061 Benchmarks
- *People can often learn about things around them by just observing those things carefully, but sometimes they can learn more by doing something to the things and noting what happens.*
- *Raise questions about the world around them and be willing to seek answers to some of them by making careful observations and trying things out.*
- *Offer reasons for their findings and consider reasons suggested by others.*

Science
Physical science
    chemistry

Integrated Processes
Observing
Comparing and contrasting
Inferring
Applying

Materials
Baking soda
Vinegar
2 quart jars
Birthday candle
Jar lid or 12cm x 12cm piece of aluminum foil
Spoon
Matches

Background Information
Human beings and animals exhale carbon dioxide. It is produced by *oxidizing* (burning) food in our bodies. The green plants take carbon dioxide from the air (air is about 0.035% carbon dioxide) and give off oxygen. The plants combine carbon dioxide with water to make their food.

CO_2 is used in baking cakes and breads. When baking powder and yeast are moistened and heated, they release carbon dioxide gas, giving off those mouth-watering aromas and making the product rise aned hold its shape. Carbon dioxide is also used to produce the fizz or sparkle in soft drinks, beer, and sparkling wines.

Carbon dioxide gas can be changed to a solid when it is cooled to -109.3° F (-78.5° C). The solid is called "dry ice" because it does not melt to form a liquid as regular ice does. It changes from the solid state back to a gas (sublimation).

Baking soda and vinegar can be used to make carbon dioxide gas – the same kind of gas that is used in commercial fire extinguishers commonly used in homes and vehicles. The commercial extinguishers spread a dry, snow-like solid that changes to gas under atmospheric pressure. The carbon dioxide gas (CO_2) which is odorless and tasteless settles over a fire and shuts off the oxygen supply that a fire must have in order to burn. It settles, or drops to a lower elevation, because it is more dense than air.

The gas used in this activity, *Pouring Carbon Dioxide Gas,* is a result of a chemical reaction. When the acetic acid from the vinegar is combined with the sodium bicarbonate in the baking soda, water and carbon dioxide are produced. The gas formed inside the jar forces the air out. When the jar is tilted, the heavier-than-air carbon dioxide flows out.

Management
1. Use the least expensive vinegar and baking soda.
2. Before mixing the solution, have the candle attached to the jar lid or foil using some of the melted wax.
3. This activity should be conducted with adult supervision.

Procedure
1. Anchor the candle to the jar lid or foil.
2. Put three spoonfuls of baking soda into one of the jars.
3. Slowly pour about 100 ml of vinegar into the jar with the baking soda.
4. After the foaming stops, tilt the jar with the solution over the second jar and hold it there for about 10 seconds. Be careful not to pour any of the solution into the second jar.
5. Light the candle and then hold the "empty" jar about 2.5 cm above the flame and slowly tilt it over the candle.

Discussion

1. What happened when the vinegar was mixed with the baking soda? Do you think that the same thing would happen if water, instead of vinegar, were used? [When the vinegar was added to the baking soda, the mixture began to fizz and bubble producing the carbon dioxide gas. (Repeat using water to demonstrate that it will not work using H_2O.)]
2. Could anything be seen entering the empty jar when the solution was tilted over it? [The gas is colorless so it cannot be seen flowing into the jar.]
3. What happened to the flame when the "empty" jar was tilted over it? [The flame flickered and died when the CO_2 was poured onto it.]
4. What might happen if more or less vinegar and baking soda were used? [Using more or less vinegar and baking soda will result in producing more or less gas.]

Extensions

1. Light several candles and space them about four inches apart. Repeat *Procedures 1-5* and see how many of the candles can be put out with the CO_2 in the jar.
2. Make a trough using a piece of cardboard about 4"x10". Mix the solution, then hold the trough so it is at about a 45° angle with the lower end near a burning candle. Tilt the jar with the solution toward the upper end of the trough. The CO_2 will flow down the trough and extinguish the flame.
3. Drop some chips of dry ice into an empty 2-liter bottle and attach a balloon over the mouth of the bottle. Observe what happens.
4. Discuss safety precautions of having a fire extinguisher in the home.

Pouring Carbon Dioxide Gas

Materials:
Baking soda
Vinegar
2 quart jars
Birthday candle
Jar lid or a piece of aluminum foil
Tablespoon
Matches

Procedure:
1. Put 3 spoons of baking soda into one of the jars.
2. Anchor the candle to the jar lid or foil.
3. Slowly pour about 200 ml of vinegar into the jar with the baking soda.
4. After the foaming stops, tilt the jar with the solution over the empty jar and hold it there for about 10 seconds. Be careful not to pour any of the solution into the jar.
5. Light the candle and then hold the empty jar about 2 1/2 cm above the flame and slowly tilt it over the candle.

Questions:
1. What happened when the vinegar was mixed with the baking soda? Do you think that the same thing would happen if water, instead of vinegar, were used?
2. Could anything be seen entering the empty jar when the solution was tilted over it?
3. What happened to the flame when the empty jar was tilted over it?
4. What might happen if more or less vinegar and baking soda were used?

Caution: This activity should be conducted with adult supervision.

Balancing Forks

Topic
Center of gravity

Key Question
Where is the center of gravity of the balancing forks?

Focus
Students will find that a lower center of gravity is easier to balance than a higher center of gravity.

Guiding Document
Project 2061 Benchmarks
- *In something that consists of many parts, the parts usually influence one another.*
- *Something may not work as well (or not at all) if a part of it is missing, broken, worn out, mismatched, or misconnected.*

Science
Physical science
  center of gravity

Integrated Processes
Observing
Generalizing
Applying

Materials
Two forks
End of a potato or cucumber
Tall glass or empty gallon can
Match stick

Background Information
　　To understand why the forks and the potato can be balanced on the side of a glass, students must know a few things about balance. All things have a *center of gravity*. This is a point where most of the mass of an object seems to be centered. The center of gravity of a round flat object, such as a dime, is directly in the center. The center of gravity of an irregularly shaped object, such as a piece of a puzzle, is often not in the center of the object. The center of gravity of some objects, such as kidney shaped objects, is not always located on the object. The center of gravity tends to move as low to the ground as possible. As it does this, the object becomes more stable and better balanced. When the center of gravity is below the balance point, the object becomes even more stable or balanced.

Management
1. You might have to adjust the forks a couple of times before you get them to balance. Don't give up because the students will be surprised at the results.
2. After the objects are balanced and the match is burning, extinguish the flame before it crosses over the edge of the glass. Gently stroke the charred wood so it drops down into the glass.

Procedure
1. Cut off the end of a potato so that it is 5 cm wide.
2. Push the forks into the potato so that they form a 45° angle with the top sides of the fork handles facing in toward the glass.
3. Push the wooden end of the match about 2.5 cm into the cut end of the potato. The head of the match should face out of the potato.
4. With the forks down below the rim of the glass, place the match stick on the edge of the glass. Try different positions until everything balances.

Discussion
1. Where is most of the mass of the potato and forks when they are balanced? [The center of gravity (where most of the mass is centered) is below the balance point on the rim of the glass.]
2. What do you think would happen if you tried to balance it with the two forks above the rim of the glass? [The forks cannot be balanced when they are positioned above the glass.]
3. What might happen if the forks were placed at an angle less than 90°?...more than 90°? [Answers will vary. Alter the angles and try to balance it.]
4. Do athletes such as football and basketball players have better balance when they are standing straight or are in a crouched position? How does this relate to the results of this activity? [Football and basketball players crouch down to lower their centers of gravity to make themselves more stable.]

Extension
　　Have students list the ways they might have seen a low center of gravity used to improve performance in everyday life. Some suggestions might include:
a. Race cars are built with the chassis as low to the road surface as possible so they maintain their balance on turns. The same is true of sport cars.
b. Tightrope walkers hold a pole to lower their center of gravity and adjust their balance.
c. Snow skiing, water skiing, and skateboarding performance is improved as a person's weight (or center of gravity) is lowered by moving to a crouching position.

Balancing Forks

Materials:
Two forks
End of a potato
Tall glass
Wooden matches

Procedure:
1. Cut off the end of a potato so that it is 5 centimeters wide.
2. Push the forks into the potato so that they form a 45 degree angle–the top side of the fork handles should face in toward the glass.
3. Anchor the wooden match in the potato by pushing it in about 2-3 cm. The head of the match should face out.
4. With the forks down below the rim of the glass, place the match on the edge of the glass. Try different postions until everything balances.
5. After the forks are balanced, light the head of the wooden match with another match.

Questions:
1. Where is most of the mass of the potato and forks when they are balanced?
2. Could you balance it if the two forks were above the edge of the glass? Explain.
3. What might happen if the forks were placed at an angle less than 90 degrees?...more than 90 degrees?
4. How do athletes such as football and basketball players better their balance? How does this relate to the results of this activity?

Balancing Clown

Topic
Center of gravity

Key Question
How can we make the clown balance on its nose?

Focus
Students will alter the clown's center of gravity to make it balance on its nose.

Guiding Document
Project 2061 Benchmarks
- *Even a good design may fail. Sometimes steps can be taken ahead of time to reduce the likelihood of failure, but it cannot be entirely eliminated.*
- *The solution to one problem may create other problems.*
- *Everything on or anywhere near the earth is pulled toward the earth's center by gravitational force.*
- *Learning means using what one already knows to make sense out of new experiences or information, not just storing the new information in one's head.*

Science
Physical science
  center of gravity

Integrated Processes
Observing
Comparing and contrasting
Generalizing
Applying

Materials
Clown
Pencil
Two coins
Masking tape
Tagboard

Background Information
   All objects have a center of gravity. The *center of gravity* is a point where all of the weight of an object seems to be centered. It is usually located at or near the center of an object. The center of gravity tends to move as low to the ground as possible.
   Scientists are not really sure what gravity is, but they have learned a great deal about what it does. For the most part, scientists agree that gravity is a force which attracts one object to another. The earth's force of gravity pulls all objects down, but since our earth is almost round, down is toward the center. Objects are pulled toward the center of the earth until they are stopped by the ground, a tabletop, a floor, or some other object. An object does not tip over easily when its center of gravity is as close to the center of the earth as it can get. Every object has a center of gravity and can be made to balance. The point at which an object balances is called the *balance point*. When the center of gravity is below the balance point, the object has more stability than when the center of gravity is above the balance point. (Stability is the ability to resist falling over.)
   Race cars are built lower to the ground than passenger cars. This lowers the car's center of gravity and gives it greater stability so it won't tip over when turning at a high rate of speed. Football players lower their bodies when blocking and tackling to give them greater stability. Lowering their bodies actually lowers their center of gravity. Tight rope walkers hold a long pole while on the rope. As the pole bends downward, the center of gravity is also lowered and the person holding the pole has a greater stability than he/she would if no pole were used.

Management
1. Use the drawing of the clown on the student activity sheet for a pattern. You can glue it to a piece of tagboard and then cut around it or copy it directly onto the tagboard. All students should have their own clown for investigation.
2. To get the clown to balance, you can stick a small piece of clay or a piece of rolled masking tape to the hands of the clown, and attach two equal masses (pennies, nickels, small washers, etc.), one on each hand.
3. Use the rubber eraser on the end of a pencil for the platform on which to balance the clown.

Procedure
1. Trace and cut an outline of the clown on a piece of tagboard.
2. Hold the clown so its nose is resting on the eraser of a pencil. Let go when you think it is balanced.
3. Let students experiment to find different ways to balance the clown. (One method is to use a piece of masking tape and attach a penny to each of the clown's hands and then repeat *Procedure 2.*)

Discussion

1. When following *Procedure 2*, where was most of the mass of the clown? Did it balance? [When attempting to balance the clown on its nose before attaching the masses, most of the clown's mass is above the balance point (its nose) so it resists being balanced.]

2. After attaching the coins, where was most of the mass? Did it balance? [After attaching the coins or washers, most of the mass is below the balance point so the clown can be balanced on its nose.]

3. What did the placement of coins on the clown do to the center of gravity? [The placement of the masses on the clown's hands lowers its center of gravity.]

4. What happens to the balance of an object or person when its center of gravity is lowered? [When a person or object lowers his/her center of gravity, his/her balance becomes more stable. This can be demonstrated by climbing a ladder. On the first step a person's balance is almost the same as if standing on the ground. As a person moves up the ladder, he/she feels less and less stable.]

Extensions

1. Have students suggest some of the ways that we can observe things adjusting their center of gravity to improve their balance. [As indicated in *Background Information*, they might suggest race cars being built with the chassis low to the ground, athletes getting in a crouched position to improve their balance when blocking, tackling, fielding a baseball, etc.]

2. Have students a long pole or board. Where is it easiest to carry? [When held at the midpoint which is its center of gravity.]

3. Repeat the experiment with the clown. Instead of attaching a mass to the hand, attach a mass to one end of a 15 cm length of thread. Secure the other end to the hand of the clown with a piece of tape. Does the clown balance better with the mass 15 cm below the hand?

4. If there is a balance beam on the playground (some have a 2"x 4" board about a foot off the ground), have students walk the beam. Put about one pound of sand in each of two plastic gallon milk containers. Tie one container to each end of a long pole (about 6 to 8 feet long) so the containers hang about two or three feet below the pole. Walk the beam again holding the pole. Adjust the pole to different heights. What happens to the balance compared to walking the beam with the pole?

Balancing Clown

Materials:
Clown
Pencil
2 coins
Masking tape

Procedure:
1. Trace and cut an outline of the clown on a piece of tagboard.
2. Hold the clown so its nose is resting on the eraser of a pencil. Let go when you think it is balanced.
3. Try different ways to get your clown to balance on its nose.

Questions:
1. When following Procedure 2, where was most of the mass of the clown? Did it balance?
2. After working with the clown, was most of the mass above or below the balance point? Did it balance?
3. To balance the clown, what needed to happen to its center of gravity?
4. What happens to the balance of an object (or person) when its center of gravity is lowered?

The Impaled Potato

Topic
Inertia

Key Question
How can you get a straw to impale a raw potato?

Focus
Students will explore the law of inertia.

Guiding Document
Project 2061 Benchmarks

- *Raise questions about the world around them and be willing to seek answers to some of them by making careful observations and trying things out.*
- *People can often learn about things around them by just observing those things carefully, but sometimes they can learn more by doing something to the things and noting what happens.*
- *Describe and compare things in terms of number, shape, texture, size, weight, color, and motion.*
- *In the absence of retarding forces such as friction, an object will keep its direction of motion and its speed. Whenever an object is seen to speed up, slow down, or change direction, it can be assumed that an unbalanced force is acting on it.*

Science
Physical science
   inertia

Integrated Processes
Observing
Predicting
Comparing and Contrasting
Generalizing
Applying

Materials
Plastic drinking straws
Potato
Plastic or styrofoam cup

Background Information
In order to understand the results of this experiment, one must consider the scientific statement of the law of inertia: *A body at rest remains at rest, and a body in motion moves in a uniformly straight course, unless some force sets on it from the outside.* One way to demonstrate how inertia works is to tie a thin piece of string to an object that weighs about one pound, such as a book, shoe, or rock. Slowly lift the object off the ground by pulling straight up on the string. Repeat the procedure by jerking the hand up quickly. If done using the right-sized string and object, the string will break when jerked quickly; but the same weight, when lifted slowly, will not break the string.

In addition to learning that a body at rest tends to stay at rest, and a body in motion tends to continue in motion, students should also understand that a force is required in both cases to change this condition– a force is needed to put a stationary object in motion, and a force is needed to stop a moving object. If friction could be eliminated, it would take as much energy to stop a moving object as it does to make the object move. The amount of energy needed to produce this work also depends on the time and distance during which the force is being exerted.

In this demonstration, the potato is at rest and the straw is the moving object. The force, or energy, that puts the straw in motion is exerted by the person holding the straw.

Procedure
1. Place the potato over the mouth of the cup.
2. Hold the straw firmly in your hand and place one end on the potato.
3. Gradually exert downward pressure on the potato.
4. Use another straw. Hold it firmly about 15 cm above the potato and quickly thrust downward.
5. Repeat *Procedure 4* but this time place your thumb over the end of the straw.

Discussion
1. What happened in *Procedure 3* when downward pressure was gradually exerted on the potato? [As downward pressure is increased, the straw will bend and not penetrate the skin of the potato.]
2. What happened when the straw was thrust downward into the potato in *Procedure 4?* [When the straw was thrust quickly downward, the laws of inertia were at work. The potato, at rest on the cup, tended to remain at rest while the straw that was in motion tended to remain in motion. The force (by the person thrusting the straw) of the straw on the potato was sufficient to penetrate and probably go all the way through it.]
3. Was there a difference when the thumb covered the end of the straw? [Covering the end of the straw with the thumb will produce the same results.]

Extensions

1. Have students discuss and list the ways that they might observe, or experience, the effects of inertia in their everyday lives. They should be able to suggest many ways, some of which will probably include the following:

   a. When riding in a car, the body is forced back against the seat as the vehicle moves forward.

   b. If an automobile comes to a sudden stop, the passengers' bodies will have a tendency to continue moving forward. In many cases, when seatbelts are not used, bodies will continue moving forward until they are stopped by the dashboard or they continue moving through the windshield.

   c. Kicking, hitting, or throwing a ball will also demonstrate the property of inertia. The ball will continue to move away from the point where the motion began until it is stopped by the friction of the air and the pull of gravity against it.

2. Ask students to bring pictures to class that show inertia in action and display them on a bulletin board entitled *Inertia in Our World.*

3. Have some students bring other inertia demonstrations to class for sharing. Provide them the opportunity to perform the experiments for the students in other classrooms.

The Impaled Potato

Materials:
Plastic straws
Potato
Plastic or styrofoam cup

Procedure:
1. Place the potato over the mouth of the cup.
2. Holding the straw firmly in your hand, place one end on the potato.
3. Gradually exert downward pressure on the potato.
4. Use another straw. Hold it firmly about 15 cm above the potato and quickly thrust downward.
5. Repeat Procedure 4, but this time place your thumb over the end of the straw.

Questions:
1. What happened in Procedure 3 when downward pressure was gradually exerted on the potato?
2. What happened when the straw was thrust downward into the potato in Procedure 4?
3. Was there a difference when the thumb covered the end of the straw?

The Bare Necessities
Flashlight

Topic
Electrical circuit

Key Question
How can you make the bulb light using a battery, a bulb, foil wire, and a clothespin?

Focus
Students will explore different ways of making an electrical circuit.

Guiding Document
Project 2061 Benchmarks
- *When parts are put together they can do things that they couldn't do by themselves.*
- *In something that consists of many parts, the parts usually influence one another.*
- *Things that give off light often also give off heat. Heat is produced by mechanical and electrical machines, and any time one thing rubs against something else.*

Science
Physical science
electricity

Integrated Processes
Observing
Comparing and contrasting
Inferring
Generalizing

Materials
D- or C-size battery
Aluminum foil
1/2 " wide masking or transparent tape
Scissors
Flashlight bulb
Wooden clothespin

Background Information
Batteries change chemical energy to electricity. In a car battery, the action of sulfuric acid on metal plates coated with oxides of lead produces electricity. The flashlight battery used in this activity is really a cell. Batteries have a combination of cells. Many dry-cell "batteries" used in flashlights have a container of zinc filled with a damp paste of ammonium chloride and manganese dioxide (the solution is called the "electrolyte") which surrounds a carbon rod. The batteries are called "dry cells" even though the paste must be slightly moist to work properly.

An electric current is produced when the free moving electrons around the metal atoms are made to move in the same general direction through a conductor. In order to produce electricity, three things are necessary: (1) a source of electrons; (2) a conducting path; and (3) a place for the electrons to go. The tendency to yield, or give up, electrons varies from one metal to another. Lithium releases its outer electron very readily. Zinc releases its two electrons less readily, and copper even less readily than zinc. When two metals are connected by a wire and immersed in an electrolyte (conducting fluid), electrons will flow through the wire toward the metal that gives up its electrons least readily. This principle is the basis for the construction of dry cells.

Light bulbs (household and flashlight bulbs) have a thin wire, called a filament, that connects two larger wires inside the bulb. When the electrons move from one of the large wires through the filament to the other large wire, the friction of the electrons rubbing against each other causes the filament to heat and become red hot. This is what causes the bulb to glow.

Procedure
1. Attach a 15 to 20 cm strip of masking or transparent tape to a piece of aluminum foil.
2. Cut the edges of the foil away from the tape and then cut the tape down the center to make two 6 to 9 inch strips about 1/4" wide.
3. Make the bulb glow using the battery, the clothespin, the bulb, and one strip of foil.
4. Make the bulb glow without using the clothespin.

Discussion
1. Which side of the strip (the foil or tape) had to touch the battery and bulb to make the bulb glow? [The foil had to touch the battery and bulb to make it glow. The aluminum acted as the conductor for the electrons.]
2. Could you make the bulb glow without using the clothespin? [It is not necessary to use the clothespin to make the bulb glow.]
3. After assembling the taped foil and the clothespin, could you make the bulb glow by using the battery in two different positions? [The bulb can be placed

at either the positive or negative end of the battery with the same results. The electrons will be conducted along the aluminum foil and through the bulb to make it glow.]

4. How can you make the bulb glow using both pieces of taped foil? [Using two pieces of foil, the bulb can be made to glow by touching one piece of foil to the base of the bulb and the other end to the battery. The second piece of foil must touch the other end of the battery and the side of the base of the bulb.]

5. What happens to the small piece of wire inside the light bulb to make it glow? [The friction caused by the movement of the electrons along the filament makes the wire red hot.]

6. What does the battery have that makes the bulb glow? [The battery has stored chemical energy that can be changed to electrical energy to make the bulb glow.]

Extensions

1. Discuss why the light bulb gives off a brighter light when used in a flashlight. (The light given off by the bulb is reflected by the silver-plated reflector in the flashlight and focused on a small area.)

2. Have someone bring a flashlight to class and examine its parts to discover why the bulb glows when the switch is moved to the "on" position. (Two pieces of metal touch and complete the circuit by allowing the electrons to move from the battery, through the switch to the bulb, and back to the battery.)

3. Ask students to suggest ways to make the bulb glow brighter.

The Bare Necessities
Flashlight

Materials:
D or C size battery (flashlight cell)
Aluminum foil
Masking or transparent tape
Scissors
Flashlight bulb
Clothespin

Procedure:
1. Attach a 15 to 20 cm strip of masking or transparent tape to a piece of aluminum foil.
2. Cut the edges of the foil away from the tape and then cut the tape down the center to make two 15 to 20 cm strips.
3. Make the bulb glow using the battery, the clothespin, the bulb, and one strip of foil.
4. Make the bulb glow without using the clothespin.

Questions:
1. Which side of the strip (the foil or tape) had to touch the battery and bulb to make the bulb glow?
2. Could you make the bulb glow without using the clothespin? Explain.
3. After assembling the taped foil and the clothespin, could you make the bulb glow by using the battery in two different positions?
4. How can you make the bulb glow using both pieces of taped foil?
5. What happens to the small piece of wire inside the light bulb to make it glow?
6. What does the battery have that makes the bulb glow?

Sandy Magnets

Topic
Magnetism

Key Question
What can we find out about sand particles by dragging a magnet through them?

Focus
Students will observe that sand often contains particles of iron.

Guiding Document
Project 2061 Benchmarks
- *People can often learn about things around them by just observing those things carefully, but sometimes they can learn more by doing something to the things and noting what happens.*
- *Raise questions about the world around them and be willing to seek answers to some of them by making careful observations and trying things out.*
- *Offer reasons for their findings and consider reasons suggested by others.*

Integrated Processes
Observing
Comparing and Contrasting
Inferring
Applying

Materials
2 paper or plastic cups
Sand
Magnet
Water
Paper
Nail

Background Information
To understand what produces a magnetic field in a permanent magnet, it is necessary to remember that all objects, including magnets, are made of atoms. Each atom consists of a positively charged nucleus surrounded by negatively charged electrons. Even when a magnet lies still, it contains atoms in which the electrons spin and move around the nucleus, somewhat like the way the earth rotates and revolves around the sun. Each spinning electron acts like a small magnet. Each pair of electrons that spins in opposite directions cancels out each others' magnetic fields. In most materials, most magnetism is canceled out in this way; in each iron atom, however, there are four electrons whose magnetic fields are not canceled. This makes each iron atom a tiny magnet. Some other elements, such as nickel, cobalt, or the rare earths, are magnetic to a lesser degree. In many common magnets, made from alloys containing these elements, the spin described above provides virtually all of the magnetic properties within the domains.

Why do magnets and magnetic materials act the way they do? The answer lies in the fact that iron and other magnetic materials are made up of very small areas called *magnetic domains*. Each domain consists of billions of atoms. The atoms within each individual domain are always perfectly aligned magnetically. In contrast, the domains themselves may or may not be aligned.

The proportion of domains that are aligned determines the amount of magnetic force a magnet has. A piece of pure iron might have almost all unaligned domains and very little magnetic strength. If a moderately strong magnet is held near this iron or strokes it, some of the iron's domains will become aligned and it will exhibit some magnetic properties. If only a few domains become aligned, the object will become slightly magnetic; if most of the domains become aligned, the material will be much more strongly magnetic. No material ever has all of its domains aligned.

The first magnets were stones (lodestones) discovered in an ancient country of Asia Minor, called Magnesia. People were puzzled by the way the stones attracted metal and many believed the magnets were "magic" stones. When a stone was hung from a string it turned until one end pointed north and the other end pointed south. This discovery led to the invention of the compass which enabled sailors to better navigate their ships because they could tell in which direction they were sailing, even in a storm.

Scientists believe that the formation of some iron ores began over a billion years ago. When great volcanoes spouted dust into the air, it fell into streams and rivers. The water dissolved iron out of the rocks and carried the iron particles into the oceans. As the iron fell to the bottom along with sand and silt it formed layers on the bottom of the oceans. Heat and pressure compressed these layers to form rock.

Over millions of years. earthquakes and glaciers brought these layers of rocks to the surface and moved them to different areas. As the ice melted, rivers formed and cut grooves in the rocks, transforming them to gravel and sand which was deposited over the surface of the earth. Today we can move a magnet through dirt and sand and the iron molecules that were formed millions of years ago will be attracted to it.

Management
1. A small magnet can be purchased at a hardware store for about 25¢.
2. Collect about 250 ml of dry sand from the yard or playground.
3. Use a large steel nail – do not use a nail made of aluminum.

Procedure
1. Move the nail back and forth through the sand and then dip it into the cup of water.
2. Move the magnet through the sand. Feel the magnet and determine if there is something clinging to it.
3. Dip the magnet into the cup of water.
4. Sprinkle some sand onto a piece of paper. Slowly move the magnet across the sand about 2.5 cm above the paper.
5. Scrape the particles clinging to the magnet onto a clean piece of paper.

Discussion
1. What happened when the nail was moved through the sand and then dipped into the water? [After the nail was moved through the sand it had some dust and sand clinging to it. This was probably removed when it was immersed in the water.]

2. What happened when the magnet was moved through the sand? [When the magnet was moved back and forth in the sand, iron particles were attracted to it.]
3. Did anything unusual happen when the magnet was dipped into the water? [When the magnet was immersed in the water, the dirt washed away and the iron was held to the magnet.]
4. Where did the stuff on the magnet come from? [The iron particles came from the iron ore that formed in the earth billions of years ago and was carried to all parts of the earth by a combination of volcanoes, earthquakes, and glaciers.]

Extensions
1. Use the information given in *Background Information* to indicate what scientists believe happened to make the dirt and sand contain iron particles.
2. Have students list the many ways that they might observe magnets being used around the home.
3. Use a world map to find out where Magnesia was located in Asia Minor.
4. Make the nail a temporary magnet by stroking it about 50 times with the magnet (stroke in same direction, not back and forth). Move the nail in the sand to see if it will attract the iron particles.

Sandy Magnets

Materials:

2 paper or plastic cups
Sand
Magnet
Water
Paper
Nail

Procedure:

1. Move the nail back and forth through the sand and then dip it into the cup of water.
2. Move the magnet through the sand. Observe the magnet and determine if there is something clinging to it.
3. Dip the magnet into the cup of water.
4. Sprinkle some sand onto a piece of paper. Slowly move the magnet across the sand about 2.5 cm above the paper.
5. Scrape the particles clinging to the magnet onto a clean piece of paper.

Questions:

1. What happened when the nail was moved through the sand and then dipped into the water?
2. What happened when the magnet was moved through the sand?
3. Did anything unusual happen when the magnet was dipped into the water?
4. Where did the "stuff" on the magnet come from?

The AIMS Program

AIMS is the acronym for "**A**ctivities **I**ntegrating **M**athematics and **S**cience." Such integration enriches learning and makes it meaningful and holistic. AIMS began as a project of Fresno Pacific College to integrate the study of mathematics and science in grades K-9, but has since expanded to include language arts, social studies, and other disciplines.

AIMS is a continuing program of the non-profit AIMS Education Foundation. It had its inception in a National Science Foundation funded program whose purpose was to explore the effectiveness of integrating mathematics and science. The project directors in cooperation with eighty elementary classroom teachers devoted two years to a thorough field-testing of the results and implications of integration.

The approach met with such positive results that the decision was made to launch a program to create instructional materials incorporating this concept. Despite the fact that thoughtful educators have long recommended an integrative approach, very little appropriate material was available in 1981 when the project began. A series of writing projects have ensued and today the AIMS Education Foundation is committed to continue the creation of new integrated activities on a permanent basis.

The AIMS program is funded through the sale of this developing series of books and proceeds from the Foundation's endowment. All net income from book and poster sales flow into a trust fund administered by the AIMS Education Foundation. Use of these funds is restricted to support of research, development, publication of new materials, and partial scholarships for classroom teachers participating in writing and field testing teams. Writers donate all their rights to the Foundation to support its on-going program. No royalties are paid to the writers.

The rationale for integration lies in the fact that science, mathematics, language arts, social studies, etc., are integrally interwoven in the real world from which it follows that they should be similarly treated in the classroom where we are preparing students to live in that world. Teachers who use the AIMS program give enthusiastic endorsement to the effectiveness of this approach.

Science encompasses the art of questioning, investigating, hypothesizing, discovering and communicating. Mathematics is the language that provides clarity, objectivity, and understanding. The language arts provide us powerful tools of communication. Many of the major contemporary societal issues stem from advancements in science and must be studied in the context of the social sciences. Therefore, it is timely that all of us take seriously a more holistic mode of educating our students. This goal motivates all who are associated with the AIMS Program. We invite you to join us in this effort.

Meaningful integration of knowledge is a major recommendation coming from the nation's professional science and mathematics associations. The American Association for the Advancement of Science in *Science for All Americans* strongly recommends the integration of mathematics, science and technology. The National Council of Teachers of Mathematics places strong emphasis on applications of mathematics such as are found in science investigations. AIMS is fully aligned with these recommendations.

Extensive field testing of AIMS investigations confirms these beneficial results.

1. Mathematics becomes more meaningful, hence more useful, when it is applied to situations that interest students.
2. The extent to which science is studied and understood is increased, with a significant economy of time, when mathematics and science are integrated.
3. There is improved quality of learning and retention, supporting the thesis that learning which is meaningful and relevant is more effective.
4. Motivation and involvement are increased dramatically as students investigate real world situations and participate actively in the process.

We invite you to become part of this classroom teacher movement by using an integrated approach to learning and sharing any suggestions you may have. The AIMS Program welcomes you!

AIMS Education Foundation Programs

A Day With AIMS

Intensive one-day workshops are offered to introduce educators to the philosophy and rationale of AIMS. Participants will discuss the methodology of AIMS and the strategies by which AIMS principles may be incorporated into curriculum. Each participant will take part in a variety of hands-on AIMS investigations to gain an understanding of such aspects as the scientific/mathematical content, classroom management, and connections with other curricular areas. The *A Day With AIMS* workshops may be offered anywhere in the United States. Necessary supplies and take-home materials are usually included in the enrollment fee.

AIMS One-Week Workshops

Throughout the nation, AIMS offers many one-week workshops each year, usually in the summer. Each workshop lasts five days and includes at least 30 hours of AIMS hands-on instruction. Participants are grouped according to the grade level(s) in which they are interested. Instructors are members of the AIMS Instructional Leadership Network. Supplies for the activities and a generous supply of take-home materials are included in the enrollment fee. Sites are selected on the basis of applications submitted by educational organizations. If chosen to host a workshop, the host agency agrees to provide specified facilities and cooperate in the promotion of the workshop. The AIMS Education Foundation supplies workshop materials as well as the travel, housing, and meals for instructors.

AIMS One-Week Fresno Pacific College Workshops

Each summer, Fresno Pacific College offers AIMS one-week workshops on the campus of Fresno Pacific College in Fresno, California. AIMS Program Directors and highly qualified members of the AIMS National Leadership Network serve as instructors.

The Science Festival and the Festival of Mathematics

Each summer, Fresno Pacific College offers a Science Festival and a Festival of Mathematics. These two-week festivals have gained national recognition as inspiring and challenging experiences, giving unique opportunities to experience hands-on mathematics and science in topical and grade level groups. Guest faculty includes some of the nation's most highly regarded mathematics and science educators. Supplies and take-home materials are included in the enrollment fee.

The AIMS Instructional Leadership Program

This is an AIMS staff development program seeking to prepare facilitators for leadership roles in science/math education in their home districts or regions. Upon successful completion of the program, trained facilitators become members of the AIMS Instructional Leadership Network, qualified to conduct AIMS workshops, teach AIMS in-service courses for college credit, and serve as AIMS consultants. Intensive training is provided in mathematics, science, processing skills, workshop management, and other relevant topics.

College Credit and Grants

Those who participate in workshops may often qualify for college credit. If the workshop takes place on the campus of Fresno Pacific College, that institution may grant appropriate credit. If the workshop takes place off-campus, arrangements can sometimes be made for credit to be granted by another college or university. In addition, the applicant's home school district is often willing to grant in-service or professional development credit. Many educators who participate in AIMS workshops are recipients of various types of educational grants, either local or national. Nationally known foundations and funding agencies have long recognized the value of AIMS mathematics and science workshops to educators. The AIMS Education Foundation encourages educators interested in attending or hosting workshops to explore the possibilities suggested above. Although the Foundation strongly supports such interest, it reminds applicants that they have the primary responsibility for fulfilling *current* requirements.

For current information regarding the programs described above, please complete the following:

Information Request

Please send current information on the items checked:

____ *Basic Information Packet* on AIMS materials
____ *Festival of Mathematics*
____ *Science Festival*
____ *AIMS Instructional Leadership Program*

____ *AIMS One-Week Fresno Pacific College Workshops*
____ *AIMS One-Week Workshops*
____ Hosting information for *A Day With AIMS* workshops
____ Hosting information for *A Week With AIMS* workshops

Name _____

Address _____
       Street   City   State   Zip

AIMS Program Publications

GRADES K-4 SERIES
Bats Incredible
Brinca de Alegria Hacia la Primavera con las Matemáticas y Ciencias
Cáete de Gusto Hacia el Otoño con la Matemáticas y Ciencias
Fall Into Math and Science
Glide Into Winter With Math and Science
Hardhatting in a Geo-World
Jaw breakers and Heart Thumpers (Revised Edition, 1995)
Overhead and Underfoot (Revised Edition, 1994)
Patine al Invierno con Matemáticas y Ciencias
Popping With Power (Revised Edition, 1994)
Primariamente Física (Revised Edition, 1994)
Primariamente Plantas
Primarily Physics (Revised Edition, 1994)
Primarily Plants
Sense-able Science
Spring Into Math and Science

GRADES K-6 SERIES
The Budding Botanist
El Botanista Principiante
Critters
Mostly Magnets
Principalmente Imanes
Ositos Nada Más
Primarily Bears
Water Precious Water

GRADES 5-9 SERIES
Down to Earth
Electrical Connections
Conexiones Eléctricas
Finding Your Bearings (Revised Edition, 1994)
Floaters and Sinkers (Revised Edition, 1995)
From Head to Toe
Fun With Foods
Historical Connections in Mathematics, Volume I
Historical Connections in Mathematics, Volume II
Historical Connections in Mathematics, Volume III
Machine Shop
Math + Science, A Solution
Our Wonderful World
Out of This World (Revised Edition, 1994)
Pieces and Patterns, A Patchwork in Math and Science
Piezas y Diseños, un Mosaic de Matemáticas y Ciencias
Soap Films and Bubbles
The Sky's the Limit (Revised Edition, 1994)
Through the Eyes of the Explorers: Minds-on Math & Mapping
Off the Wall Science: A Poster Series Revisited

FOR FURTHER INFORMATION WRITE TO:
AIMS Education Foundation • P.O. Box 8120 • Fresno, California 93747-8120

We invite you to subscribe to $\mathcal{AIMS}$!

Each issue of $\mathcal{AIMS}$ contains a variety of material useful to educators at all grade levels. Feature articles of lasting value deal with topics such as mathematical or science concepts, curriculum, assessment, the teaching of processing skills, and historical background. Several of the latest AIMS math/science investigations are always included, along with their reproducible activity sheets. As needs direct and space allows, various issues contain news of current developments, such as workshop schedules, activities of the AIMS Instructional Leadership Network, and announcements of upcoming publications.

$\mathcal{AIMS}$ is published monthly, August through May. Subscriptions are on an annual basis only. A subscription entered at any time will begin with the next issue, but will also include the previous issues of that volume. Readers have preferred this arrangement because articles and activities within an annual volume are often interrelated.

Please note that an $\mathcal{AIMS}$ subscription automatically includes duplication rights for one school site for all issues included in the subscription. Many schools build cost-effective library resources with their subscriptions.

YES! I am interested in subscribing to $\mathcal{AIMS}$.

Name _____   Home Phone _____

Address _____   City, State, Zip _____

Please send the following volumes (subject to availability):

_____	Volume I (1986-87)	$27.50	_____	Volume VI (1991-92)	$27.50
_____	Volume II (1987-88)	$27.50	_____	Volume VII (1992-93)	$27.50
_____	Volume III (1988-89)	$27.50	_____	Volume VIII (1993-94)	$27.50
_____	Volume IV (1989-90)	$27.50	_____	Volume IX (1994-95)	$27.50
_____	Volume V (1990-91)	$27.50	_____	Volume X (1995-96)	$27.50

_____ Limited offer: Volumes IX & X (1994-95 & 1995-96) $50.00

(Note: Prices may change without notice. For current prices, call (209) 255-4094.)

Check your method of payment:

☐ Check enclosed in the amount of $ _____

☐ Purchase order attached (Please be sure it includes the P.O. number, the authorizing signature, and the position of the authorizing person.)

☐ Credit Card   (Check One)
☐ Visa   ☐ MasterCard   Number _____

Amount $ _____   Expiration Date _____

Signature _____   Today's Date _____

Make checks payable to **AIMS Education Foundation**.
Mail to $\mathcal{AIMS}$ magazine, **P.O. Box 8120, Fresno, CA 93747-8120.**

AIMS Duplication Rights Program

AIMS has received many requests from school districts for the purchase of unlimited duplication rights to AIMS materials. In response, the AIMS Education Foundation has formulated the program outlined below. There is a built-in flexibility which, we trust, will provide for those who use AIMS materials extensively to purchase such rights for either individual activities or entire books.

It is the goal of the AIMS Education Foundation to make its materials and programs available at reasonable cost. All income from sale of publications and duplication rights is used to support AIMS programs. Hence, strict adherence to regulations governing duplication is essential. Duplication of AIMS materials beyond limits set by copyright laws and those specified below is strictly forbidden.

Limited Duplication Rights

Any purchaser of an AIMS book may make up to *200 copies* of any activity in that book for use at *one school site*. Beyond that, rights must be purchased according to the appropriate category.

Unlimited Duplication Rights for Single Activities

An individual or school may purchase the right to make an unlimited number of copies of a single activity. The royalty is $5.00 per activity per school site.

Examples: 3 activities x 1 site x $5.00 = $15.00
9 activities x 3 sites x $5.00 = $135.00

Unlimited Duplication Rights for Whole Books

A school or district may purchase the right to make an unlimited number of copies of a single, *specified* book. The royalty is $20.00 per book per school site. This is in addition to the cost of the book.

Examples: 5 books x 1 site x $20.00 = $100.00
12 books x 10 sites x $20.00 = $2400.00

Magazine/Newsletter Duplication Rights

Members of the AIMS Education Foundation who Purchase the *AIMS* magazine/*Newsletter* are hereby granted permission to make up to 200 copies of any portion of it, provided these copies will be used for educational purposes.

Workshop Instructors' Duplication Rights

Workshop instructors may distribute to registered workshop participants a maximum of 100 copies of any article and /or 100 copies of no more than 8 activities, provided these six conditions are met:

1. Since all AIMS activities are based upon the *AIMS Model of Mathematics* and the *AIMS Model of Learning*, leaders must include in their presentations an explanation of these two models.
2. Workshop instructors must relate the AIMS activities presented to these basic explanations of the AIMS philosophy of education.
3. The copyright notice must appear on all materials distributed.
4. Instructors must provide information enabling participants to apply for membership in the AIMS Education Foundation or order books from the Foundation.
5. Instructors must inform participants of their limited duplication rights as outlined below.
6. Only student pages may be duplicated.

Written permission must be obtained for duplication beyond the limits listed above. Additional royalty payments may be required.

Workshop Participants' Rights

Those enrolled in workshops in which AIMS student activity sheets are distributed may duplicate a maximum of 35 copies or enough to use the lessons one time with one class, whichever is less. Beyond that, rights must be purchased according to the appropriate category.

Application for Duplication Rights

The purchasing agency or individual must clearly specify the following:
1. Name, address, and telephone number
2. Titles of the books for Unlimited Duplication Rights contracts
3. Titles of activities for Unlimited Duplication Rights contracts
4. Names and addresses of school sites for which duplication rights are being purchased

NOTE: Books to be duplicated must be purchased separately and are not included in the contract for Unlimited Duplication Rights.

The requested duplication rights are automatically authorized when proper payment is received, although a *Certificate of Duplication Rights* will be issued when the application is processed.

Address all correspondence to

**Contract Division
AIMS Education Foundation
P.O. Box 8120
Fresno, CA 93747-8120**